T0214334

SpringerBriefs in Physics

More information about this series at http://www.springer.com/series/8902

Jarosław Pykacz

Quantum Physics, Fuzzy Sets and Logic

Steps Towards a Many-Valued
Interpretation of Quantum Mechanics

 Springer

Jarosław Pykacz
Institute of Mathematics
University of Gdańsk
Gdańsk
Poland

ISSN 2191-5423   ISSN 2191-5431   (electronic)
SpringerBriefs in Physics
ISBN 978-3-319-19383-0   ISBN 978-3-319-19384-7   (eBook)
DOI 10.1007/978-3-319-19384-7

Library of Congress Control Number: 2015940987

Springer Cham Heidelberg New York Dordrecht London

Printed on acid-free paper

Springer International Publishing AG Switzerland is part of Springer Science+Business Media
(www.springer.com)

Contents

Chapter 1
Introduction

After von Weizsäcker's work no serious attempt seems ever to have been made to elaborate further a many-valued logical approach to quantum mechanics
(Max Jammer, The Philosophy of Quantum Mechanics, (1974), p. 379)

Quantum mechanics is an extremely efficient theory and up-to-now no experiment designed to check it yielded results indicating that it could be wrong. Even more: present technological progress approaches the point at which it will be possible to gain full control on single quanta, which would make marvellous predictions of two newly-born branches of quantum mechanics: quantum information and quantum computation [1] fully realizable. However, this theoretical and technological progress is not accompanied by the progress in our "understanding" (whatever it could mean) of quantum phenomena.

According to Webster's Third New International Dictionary "interpretation" means "explanation of what is not immediately plain or explicit". Indeed, quantum mechanics is full of concepts, symbols and objects that are not immediately plain or explicit since they have no counterparts in our everyday life. Actually, how could we imagine, for example, a material object being simultaneously in two distinct places or being at the same time a particle and a wave, when all our "macroscopic" experience says that this is impossible? An "interpretation" of quantum mechanics should explain at least some of such conundrums in a way that could be accepted by us: macroscopic beings whose intuition grows up exclusively on macroscopic phenomena.

The interpretation of quantum mechanics proposed in this work is based on many-valued logic and observation that people, usually unconsciously, use this logic while considering future events, the occurence or non-occurence of which is not sure at present. Jan Łukasiewicz formalized this idea in his numerous papers [2], and

© The Author(s) 2015
J. Pykacz, *Quantum Physics, Fuzzy Sets and Logic*,
SpringerBriefs in Physics, DOI 10.1007/978-3-319-19384-7_1

argued that truth-values of non-certain statements concerning future events (*future contingents*) equal the probability (possibility? likelihood?) that these statements will, at due time, occur to be true. In "macroscopic" cases considered by Łukasiewicz these probabilities were supposed to be evaluated in a subjective and imprecise way. Fortunately, in quantum mechanics these probabilities are provided by the theory and are precisely known.

Quantum physics and modern theory of many-valued logics were born nearly simultaneously in the third decade of the twentieth century. However, the attempts at applying many-valued logics to the description of quantum systems expired soon after World War II. This was the situation that persisted at least until the early seventies when Max Jammer in his book [3] published in 1974 wrote the words which have been chosen as a kind of *anti*-motto to this work.

The recently observed revival of interest in applying many-valued logics to the description of quantum phenomena is closely connected with a new and rapidly developing branch of mathematics: fuzzy set theory. Fuzzy sets remain in the same relation to infinite-valued logic as traditional sets to classical two-valued logic. Therefore, truth-values of many-valued statements about results of future experiments on quantum objects may be, equivalently, treated as degrees to which these objects possess respective properties before they are measured.

The book is organized as follows: A brief survey of main interpretations of quantum mechanics is given in Chap. 2. Chapter 3 contains introduction to many-valued logics while Chap. 4 gives the rudiments of the fuzzy set theory and shows its links with the infinite-valued Łukasiewicz logic. Chapter 5 contains a short historical survey of attempts at applying non-classical logics to the description of quantum phenomena, from Zawirski's attempts in the early thirties to von Weizsäcker's papers published in the late fifties of the twentieth century. Out of these attempts only the Birkhoff and von Neumann proposal to use a two-valued but non-distributive logic gained wide popularity and is still in use nowadays. Chapter 6, of rather technical character, is devoted to this kind of "quantum logic" and presents it through three models: the traditional algebraic model, Mączyński's functional model, and finally the fuzzy set one, elaborated in a series of papers by the present author. The fuzzy set model of the Birkhoff–von Neuman quantum logic enables it to be expressed immediately in the language of the infinite-valued Łukasiewicz logic. This procedure, developed in Chap. 7, allows the Birkhoff–von Neuman quantum logic to be treated as a special kind of infinite-valued Łukasiewicz's logic with partially defined conjunction and disjunction. This unifies two competing approaches: the many-valued, and the two-valued but non-distributive one, which have co-existed in the quantum logic theory since its very beginning. This also clarifies the long-standing problem of proper models for the disjunction and conjunction of experimentally verifiable propositions about quantum systems and allows a logical analysis to be performed of the two-slit experiment.

Chapter 8 contains some speculations about the new perspectives opened by the proposed approach. Finally, Chap. 9 is devoted to the concise exposition of the proposed many-valued interpretation of quantum mechanics, performed in a way similar to the way in which other interpretations of quantum mechanics were presented in Chap. 2, which makes their comparison more easy.

This book was financed by the grant 2011/03/B/HS1/04573 of the Polish National Foundation for Science (NCN).

References

1. Nielsen, M. A. and I. L. Chuang, *Quantum Computation and Quantum Information* (Cambridge University Press, Cambridge, 2000).
2. Łukasiewicz, J. *Selected Works*, ed. by L. Borkowski (North-Holland, Amsterdam, and PWN—Polish Scientific Publishers, Warszawa, 1970).
3. Jammer, M. *The Philosophy of Quantum Mechanics* (Wiley-Interscience, New York, 1974).

Chapter 2
A Brief Survey of Main Interpretations of Quantum Mechanics

Since descriptions and comments on the plethora of various interpretations of quantum mechanics are widely accessible (see, e.g., [1, 2]), we give in this chapter only a very brief survey of the most popular of them. We stress that our presentation and evaluation of various interpretations is highly subjective. In particular, in our opinion ontic determinism precludes the existence of free will, which we treasure, therefore indeterminism is in our opinion a virtue, not a drawback of an interpretation.

The other difficulty in presenting such a brief survey of various interpretations of QM is caused by the fact that most of them are not uniquely defined. We tried in each case to extract a bunch of ideas that could be treated as a "common denominator" by various adherents of an interpretation, but in many cases this occurred to be a difficult task.

It should be also noticed that not all "interpretations of QM" presented in the literature are interpretations in the strict sense of this word, i.e., interpretations of the "bare" mathematical (Hilbert space) formalism of the orthodox quantum theory. In many cases an "interpretation" introduces or at least foresees various modifications of the usual mathematical formalism of QM, so it should be rather called a "theory". Since in this brief survey we decided to confine to "interpretations of QM" in the strict sense of this word, we do not mention here such important proposals as Ghirardi et al. [3], or other "Objective Collapse Theories", or "Hidden Variables Interpretations".

The simplest test that allows to distinguish between an *interpretation* and a *theory* is the existence or nonexistence of experimental proposals that could, at least theoretically, distinguish it from the other ones, since no two interpretations of QM, by the very definition of this notion, could be distinguished in this way. Therefore, if a set of ideas pertaining to QM allows for its experimental discrimination from the other ones, it should be rather called a *theory*, not an *interpretation*. However, in many cases this issue is not settled even among various proponents of a specific interpretation, which causes the issue of filtering out *interpretations* from *theories* an extremely difficult task.

© The Author(s) 2015
J. Pykacz, *Quantum Physics, Fuzzy Sets and Logic*,
SpringerBriefs in Physics, DOI 10.1007/978-3-319-19384-7_2

2.1 Ensemble Interpretation

Ensemble Interpretation (EI), called also Statistical Interpretation, takes literally Born's probabilistic interpretation of squared modulus of the wave function. Therefore, it assumes that the wave function does not refer to an individual quantum object, but to a statistical ensemble of such "identically prepared" objects. This ensemble can be either meant literally, as it is in the case of myriads of identically prepared photons emitted by a source, or it can be meant "abstractly" as an "imaginary collection" of multiple copies of an individual object. It seems that this interpretation of QM was supported by Einstein who, however, went further and inferred from it that the "orthodox" QM should be supplemented by hidden variables, while in general there is no such assumption in contemporary expositions of the EI. More recently EI was promoted vigorously by Ballentine [4, 5] (see also extensive bibliography at Ulf Klein's website [6]).

Main idea:

- Wave function is an abstract concept that refers to an ensemble of quantum systems. In particular, there does not exist anything like "wave function of an individual quantum system".

Virtues:

- EI is "minimal" in the sense that it does not make use of any metaphysical assumptions.
- No problems with measurements, collapse, Schrödinger's cats, etc.

Drawbacks:

- EI does not satisfy our deep desire for "final answers".
- Impossibility to explain "quantum Zeno effect".

2.2 Copenhagen Interpretation

Out of all interpretations of quantum mechanics proposed up to now, the Copenhagen Interpretation (CI), in spite of being still the most popular (see the results of a poll executed by Schlosshauer et al. [7]), is the worst-defined one. According to Peres [8]: "There seems to be at least as many different Copenhagen Interpretations as people who use that term, probably there are more".

CI has its roots in Bohr's and Heisenberg's ideas elaborated in the town of Copenhagen in the late twenties of the XX century. Nevertheless, the very name "Copenhagen Interpretation" was attached to this bunch of ideas not before than in the fifties. It should be also noticed that ideas usually presented in textbooks as CI are not entirely identical with original ideas of Bohr and Heisenberg which, moreover, were also different from each other in some details.

Main ideas:

- Quantum objects display either wave-like or particle-like properties. It is an experimental arrangement that defines which properties can be observed.
- Quantum mechanics is fundamentally about observations or results of measurements.
- It is meaningless to talk about properties of quantum objects before they are measured.
- Wave function is a mathematical concept. Physical meaning has its squared modulus which, according to Born's rule, defines probabilities of obtaining various experimental results.
- Wave functions evolve in two ways:

  1. Deterministically, according to Schrödinger equation, when no measurement is made.
  2. Indeterministically ("collapse" or "reduction") when measurement is made.

- Hilbert space description of quantum phenomena is the ultimate one. In particular, there are no hidden variables that could explain random behaviour of quantum objects. Therefore, quantum probabilities are ontic, not epistemic.

Virtues:

- Fundamental indeterminism of the quantum world.

Drawbacks:

- Artificial division of the physical world into the quantum world and the classical world.
- The "objectification problem", i.e., a problem how "potential" properties become "actual" in the course of a measurement.

2.3 Pilot-Wave Interpretation

The Pilot-Wave Interpretation (PWI), known also as Causal or Ontological Interpretation, de Broglie–Bohm theory, or Bohmian mechanics, is based on the ideas presented by de Broglie in 1927 in a paper [9] published in *Le Journal de Physique et le Radium* and also presented at the 5th Solvay Conference, and later on rediscovered by Bohm [10]. It seems that the majority of advocates of this interpretation (although not all) maintain that all experimental predictions of the de Broglie–Bohm theory are exactly the same as predictions of the "orthodox" QM, therefore according to them, it is really an *interpretation* of QM in the narrow sense of this word.

Main ideas:

- Both "wave-like" and "particle-like" aspects of quantum objects have simultaneous reality: quantum particles move along definite trajectories guided by their pilot waves. In particular, in a two-slit experiment a particle goes through one slit only but its pilot wave goes through both slits, interferes with itself, and attracts the particle to the areas of constructive interference.

- Pilot waves are represented mathematically by solutions of Schrödinger equation. They never collapse.
- The actual positions of particles are "hidden variables".

Virtues:

- PWI provides a "classical-like", visible and easy to comprehend image of the microworld.
- No measurement problem.

Drawbacks:

- Manifest nonlocality.
- Determinism.

2.4 Many-Worlds Interpretation

The cornerstone of the Many Worlds Interpretation (MWI) was laid down by Hugh Everett III in his PhD thesis [11] (reprinted in [12], see also paper [13] based on this thesis). Nevertheless, it should be noticed that Everett himself never jumped into far-reaching ontological conclusions drawn by his followers, and only stated enigmatically: "*From the present viewpoint all elements of superposition are equally 'real'*" ([12], pp. 116–117).

Actually, the very name MWI and explicit formulation of the idea that "*every quantum transition taking place on every star, in every galaxy, in every remote corner of the universe is splitting our local world on earth into myriads of copies of itself*" is due to DeWitt [14].

Among other distinguished advocates of the MWI are Deutsch [15, 16] and Vaidman [17]. It should be noticed that according to the results of a poll executed by Schlosshauer et al. [7]), the MWI occurred to be the second w.r.t. popularity after the Copenhagen Interpretation.

Main ideas:

- There exists the "basic physical entity": the universal wave function, that never collapses.
- At every "moment of choice": a photon either passes through a semi-transparent mirror or is reflected, Schrödinger's cat is either poisoned or not, a universe that we witness (which is only one copy of myriads of its copies that form the *Multiverse*) splits into separate, equally real copies in which either this or that course of events takes place. Adherents of the MWI are not unanimous whether these different copies can somehow "influence" or "feel the existence" of the others or not.

Virtues:

- Observers and measurements play no special role.
- No problems with collapse.
- According to Vaidman [17] *"The MWI resolves most, if not all, paradoxes of quantum mechanics."*

Drawbacks:

- Extremely weird ontology.
- The very idea of replacing the unique Universe by myriads of its copies that form the *Multiverse* seems to be in deep contradiction to the idea of the Ockham Razor that successfully guides Western Philosophy for centuries.
- Indeterminism observed in the microworld is only apparent since the universal wave function evolves deterministically.

2.5 Consistent Histories Interpretation

The Consistent Histories Interpretation (CHI) is sometimes proclaimed by its advocates as "Copenhagen done right". It was originated by Griffiths [18, 19], followed by Omnès [20, 21], and by Gell-Mann and Hartle [22] who used the term "decoherent histories". It is based on the notion of a *history* which is thought of as a time-sequence of properties actually possessed by a quantum object in consecutive instants of time. This sequence is mathematically represented by a tensor product of projection operators. Boundles of such histories, called *frameworks* are analogs of sample spaces in classical probability theory, and allow to define on them probabilities that coincide with probabilities yielded by Born's rule. However, it should be stressed that to a specific *framework* belong only histories that are *consistent* in the sense that at any instant of time they do not contain properties represented by non-commuting projectors.

Main ideas:

- Wave function is a tool for calculating probabilities, not a representation of reality.
- Time development of a quantum system is a stochastic process.
- All frameworks are equally "real".
- The *single framework rule*: Any discussion about properties of quantum objects has to be confined to a single framework. Using in the discussion properties that belong to incompatible frameworks is the source of paradoxes.
- Measurements reveal actually existing properties of quantum objects, however a property that exists in some frameworks may not exist in others.

Virtues:

- No measurement problem, no superluminal influences, no paradoxes.
- Fundamental indeterminism.

Drawbacks:

- Highly unclear ontology.
- Actuality of properties depends on the chosen framework ("relativity of reality").

2.6 Modal Interpretations

The name of this class of interpretations refers to modal logic, i.e., logic capable of taking into consideration sentences expressing necessity, possibility and contingency.

Originally there was a single modal interpretation (MI) of non-relativistic quantum mechanics proposed by van Fraassen [23]. Later on various researchers involved in this line of investigation developed slightly different approaches which, however, are usually collectively called "modal interpretations".

Characteristic to all MIs is a distinction between the *dynamical state* of a quantum system, which determines what *may* be the case and is just the quantum state of the orthodox QM, and the *value state* which represents all properties that the system actually possesses at a given instant. In various versions of MIs various observables are chosen as "privileged", i.e., always possessing definite values.

Main ideas:

- The standard formalism of QM, however without the projection postulate, is accepted.
- Quantum systems possess all the time definite properties that define their value states.
- The dynamical state that always evolves according to Schrödinger equation and never collapses defines what the possible properties of a system and their corresponding probabilities are.

Virtues:

- No measurement problem.
- Indeterminism.

Drawbacks:

- Unclear ontology which is, moreover, different in different versions of MIs.

2.7 Relational Quantum Mechanics

The main assumption of Relational Quantum Mechanics (RQM), originated by Rovelli [24], states that QM is not an "absolute" description of reality but rather deals with relations between various objects. Consequently, the notion of "observer-independent" description of the world is declared as being unphysical. Different observers may give different descriptions of the same event. However, it should be noticed that this refers only to "hierarchical" sets of observers: the "prime" observer is O that observes what's going on in an observed system S, the "secondary" observer is P that observes what's going on in a system $S + O$, and so on...

Main ideas:

- All physical systems are, fundamentally, quantum systems.
- QM is a "complete" theory: there are no hidden variables or other items that should be added to it.
- QM is not about properties of objects, but about relations between objects.
- Measurement is an ordinary physical interaction.
- "Absolute" or "observer-independent" state of a quantum system has no meaning.

Virtues:

- Ontological parsimony.
- It is claimed [25] that RQM allows for such reformulation of the original EPR conditions, that apparent conflict between QM and special relativity disappears.

Drawbacks:

- Relativity of properties of physical objects (even if only w.r.t. "hierarchical" set of observers).
- Not clearly stated position w.r.t. the determinism/indeterminism issue.

2.8 Other, Less Popular Interpretations

Seven main interpretations outlined above definitely do not exhaust the list of up to now proposed interpretations of QM. Among the other ones we can mention the following:

- "*Consciousness Causes Collapse*": a rather extreme point of view ascribed to von Neumann [26] and Wigner [27, 28].
- *Many Minds Interpretation* [29, 30]: a "subjective offspring" of MWI, in which the multitude of "parallel universes" is replaced by the multitude of "minds" associated with each sentient being.
- *Transactional Interpretation* [31] in which a quantum event is a result of a "transaction" between advanced (backward-in-time) and retarded (forward-in-time) waves.
- *Information Interpretation* which assumes that "the QM-formalism describes information about micro systems extracted by means of macroscopic measurement devices" [32]. This relatively new interpretation quickly gains popularity and most probably will be considered as belonging to the mainstream soon (see, e.g., [33, 34]).

2.9 Summary

All interpretations of QM presented in this Chapter are based on 2-valued logic.[1] This is not a surprise, taking into account that 2-valued logic successfully guided Western Science for centuries. Actually, till Łukasiewicz there were no alternatives, and even later on many-valued logics wandered on the fringes of the mainstream of Science, and were regarded as a mathematical curiosity with no relation to the physical world.

Most probably to the majority of scientists the idea of going beyond the 2-valued logic in the description of the physical reality is as aberrant as it would be the idea of abandoning Ptolemaic system before the Copernicus or leaving the domain of Euclidean flat geometry before Einstein.

However, the accumulation of "paradoxes" and development of more and more weird interpretations of QM is maybe a sign that this Gordian knot should be cut by transgressing the boundaries encircled by the 2-valued logic. The rest of this work is devoted to the presentation and justification of this proposal.

References

1. *The Stanford Encyclopedia of Philosophy* (Winter 2014 Edition), Edward N. Zalta (ed.), http://plato.stanford.edu/search/searcher.py?query=Interpretation+of+quantum+mechanics.
2. Wikipedia contributors, "Interpretations of quantum mechanics," *Wikipedia, The Free Encyclopedia*, http://en.wikipedia.org/w/index.php?title=Interpretations_of_quantum_mechanics&oldid=623968383 (accessed June 5, 2014).
3. Ghirardi, G. C., A. Rimini, and T. Weber, "A Model for a Unified Quantum Description of Macroscopic and Microscopic Systems", in: L. Accardi et al. (eds) *Quantum Probability and Applications* (Springer, Berlin, 1985).
4. Ballentine, L. E. "The statistical interpretation of quantum mechanics", *Reviews of Modern Physics*, **42** (1970) 358–381.
5. Ballentine, L. E. *Quantum Mechanics: A Modern Development* (World Scientific., Singapore, 1998).
6. Klein, U. *The Statistical Interpretation of Quantum Theory*, version 02, 11. 12. 2012, statintquant.net/siq.html#siqli1.html.
7. Schlosshauer, M., J. Kofler, and A. Zeilinger, "A snapshot of foundational attitudes toward quantum mechanics", arXiv: 1301.1069 [quant-ph].
8. Peres, A. "Karl Popper and the Copenhagen interpretation", *Studies in History and Philosophy of Modern Physics*, **33** (2002) 23–34.
9. de Broglie, L. "Wave mechanics and the atomic structure of matter and of radiation", *Le Journal de Physique et le Radium*, **8** (1927) 225.
10. Bohm, D. "A suggested interpretation of the quantum theory in terms of hidden variables, I" *Physical Review*, **85** (1952) 166–179; and II, **85** (1952) 180–193.
11. Everett, H. *On the Foundations of Quantum Mechanics* (PhD Thesis, Princeton University, 1957). Available online as "The Theory of the Universal Wavefunction" at http://www-tc.pbs.org/wgbh/nova/manyworlds/pdf/dissertation.pdf

---

[1] Even modal logic, which is a base of modal interpretations, although non-classical and sometimes regarded as a "coarse graining" of many-valued logics, is generally considered as 2-valued logic.

12. DeWitt, B., and R. N. Graham (eds), *The Many-Worlds Interpretation of Quantum Mechanics* (Princeton University Press, Princeton, 1973).

13. Everett, H "Relative state formulation of quantum mechanics". *Reviews of Modern Physics*, **29** (1957) 454–462.

14. DeWitt, B. "Quantum mechanics and reality: Could the solution to the dilemma of indeterminism be a universe in which all possible outcomes of an experiment actually occur?", *Physics Today*, **23** (1970) 30–40.

15. Deutsch, D. "Three experimental implications of the Everett interpretation", in: R. Penrose and C. J. Isham (eds) *Quantum Concepts of Space and Time* (Clarendon Press, Oxford, 1986) 204–214.

16. Deutsch, D. *The Fabric of Reality: The Science of Parallel Universes and Its Implications* (Penguin Books, London, 1998).

17. Vaidman, L. "Many-Worlds Interpretation of Quantum Mechanics", *The Stanford Encyclopedia of Philosophy* (Winter 2014 Edition), Edward N. Zalta (ed.), http://plato.stanford.edu/archives/win2014/entries/qm-manyworlds/.

18. Griffiths, R. "Consistent histories and the interpretation of quantum mechanics", *Journal of Statistical Physics*, **36** (1984) 219–272.

19. Griffiths, R. *Consistent Quantum Theory* (Cambridge University Press, Cambridge, 2002).

20. Omnès, R. "Logical reformulation of quantum mechanics I. Foundations", *Journal of Statistical Physics*, **53** (1988) 893–932.

21. Omnès, R. *Understanding Quantum Mechanics* (Princeton University Press, Princeton, 1999).

22. Gell-Mann, M. and J. Hartle, "Quantum mechanics in the light of quantum cosmology", in: W. Zurek (ed) *Complexity, Entropy, and the Physics of Information* (Addison-Wesley, Reading, Mass., 1990) 425–458.

23. van Fraassen, B. C. "A formal approach to the philosophy of science", in: R. Colodny (ed) *Paradigms and Paradoxes: The Philosophical Challenge of the Quantum Domain* (University of Pittsburgh Press, Pittsburgh, 1972) 303–366.

24. Rovelli, C. "Relational quantum mechanics", *International Journal of Theoretical Physics*, **35** (1996) 1637–1678.

25. Laudisa, F. "The EPR argument in a relational interpretation of quantum mechanics", *Foundations of Physics Letters*, **14** (2001) 119–132.

26. von Neuman, J. *Mathematical Foundations of Quantum Mechanics* (Princeton University Press, Princeton, 1955).

27. Wigner, E. and H. Margenau, "Remarks on the mind body question", *American Journal of Physics*, **35** (1967) 1169–1170.

28. Esfeld, M. "Essay review. Wigner's view of physical reality", *Studies in History and Philosophy of Modern Physics*, **30B** (1999) 145–154.

29. Zeh, H. D. "On the interpretation of measurements in quantum theory", *Foundations of Physics*, **1** (1970) 69–76.

30. Albert, D. and B. Loewer, "Interpreting the many worlds interpretation", *Synthese*, **77** (1988) 195–213.

31. Cramer, J. "The transactional interpretation of quantum mechanics", *Reviews of Modern Physics*, **58** (1986) 647–688.

32. Khrennikov, A. "Växjö interpretation of wave function: 2012", in: A. Khrennikov et al. (eds) *Quantum Theory: Reconsiderations of Foundations 6,* AIP Conference Proceedings **1508** (2012) 244–252.

33. Fuchs, C. "Quantum mechanics as quantum information (and only a little more)" in: A. Khrennikov (ed) *Quantum Theory: Reconsideration of Foundations* (Växjö University Press, Växjö, 2002) 463–543.

34. Zeilinger, A. *Dance of the Photons: From Einstein to Quantum Teleportation* (Farrar, Straus and Giroux, New York, 2010).

Chapter 3
A Brief Survey of Many-Valued Logics

Classical, two-valued logic deals exclusively with statements which can be unambiguously classified as being either true or false. Other statements simply do not belong to the domain of classical logic. In particular this applies to statements concerning future events, a problem already noticed by Aristotle in considering the statement since then widely quoted, "There will be a sea battle tomorrow". It should be mentioned that, despite the tradition of calling classical two-valued logic "Aristotelian logic", there are indications [1] that Aristotle himself classified "future contingents", i.e. statements about future events which are not yet decided, as neither true nor false.[1]

The problem of future contingents was also discussed in the Middle Ages and it seems that such scholars as Duns Scotus and William of Ockham in the thirteenth and the fourteenth centuries and Peter de Rivo in the fifteenth century considered such statements as indeterminate.

Modern attempts at establishing non-classical logical systems, mostly three-valued ones, began at the end of the nineteenth century. In 1897 Hugh MacColl investigated so called "three-dimensional logic" and in 1909 Charles Peirce considered "triadic logic" as a possible basis for "trichotomic mathematics". In 1910 Nicolai Vasil'ev in Kazan, Russia, built a system of three-valued "imaginary (non-Aristotelian) logic", whose name obviously referred to "imaginary (non-Euclidean) geometry" which had been presented for the first time at the same university 84 years earlier by Nicolai Lobachevskij.

Jan Łukasiewicz is generally recognized as a founding father of the modern theory of many-valued logics and his numerous papers on this subject, published from

---

[1]Łukasiewicz in many of his papers [2] claimed that the law of bivalence is actually due to the Stoics, especially Chrisippus who "...appears to have been the first logician to consciously set up and stubbornly defend the theorem that every proposition is either true or false" (quotation from [1]). Therefore, Łukasiewicz proposed to call his many-valued logic "non-Chrisippean" rather than "non-Aristotelian".

© The Author(s) 2015
J. Pykacz, *Quantum Physics, Fuzzy Sets and Logic*,
SpringerBriefs in Physics, DOI 10.1007/978-3-319-19384-7_3

1920 until his death in 1956, are well-known. In contrast to these papers, his booklet "Die logischen Grundlagen der Wahrscheinlichkeitsrechnung" [3], published in 1913, although evaluated as "one of Łukasiewicz's most valuable works" in the foreword to Łukasiewicz's *Selected Works* [2], is relatively less-known. In this booklet he considered statements containing a variable, e.g. "*x is an Englishman*" and he attributed to them truth-value equal to the ratio of the number of values of a variable for which this statement is true to the total number of values of this variable. Since he assumed the total number of values of a variable to be finite, the logic thus obtained is n-valued with n being a natural number depending on the particular situation described by a proposition. Łukasiewicz's principal aim in his 1913 paper [3] was to give the logical background to the notion of probability which at that time was much more alien to the rest of mathematics than it is now. An n-valued non-classical logic, which nowadays can be classified as a *probability logic* was only a kind of a by-product of these efforts and never gained such popularity as his later versions of many-valued logics, which were investigated after 1920.

The year 1920 is generally recognized as the year of the birth of the modern theory of many-valued logics. In fact, in this year two seminal papers on this theory were published independently by Jan Łukasiewicz in Poland [4] and by Emil Post in the USA [5]. Łukasiewicz arrived at his construction of a three-valued logic after a long period of philosophical investigations concerning the problems of determinism (cf. his numerous papers collected in [2], especially [6]), and of modal propositions, i.e. propositions of the form: "*It is possible (impossible, contingent, necessary) that…*" [1]. He openly declared himself a devoted adherent of indeterminism who, in his own words [7], "*…declared a spiritual war upon all coercion that restricts man's free creative activity*". The Chrisippean law of bivalence, which states that every proposition is necessarily either true or false, took the form of a fortress in this war, which had to be blown up since it blocked the way towards indeterminism. Łukasiewicz argued that determinism follows necessarily from the law of bivalence, not from the law of excluded middle, which only states that the disjunction of any proposition and its negation, e.g. "*there will be a sea battle tomorrow* OR *there will not be a sea battle tomorrow*" is always a true proposition. According to Łukasiewicz, who also claimed that this had been the original position taken by Aristotle, such disjunction may remain true even if neither of its constituents are either true or false.

Łukasiewicz in the majority of his papers on three-valued logic denoted this additional truth-value by the number $1/2$. It is possible that in the beginning he did this simply because $1/2$ lies between 0 and 1, which are the generally accepted symbols of *falsehood* and *truth*, but later on this choice turned out to be very fortunate, since it made the generalization of his three-valued logic to an n-valued or infinite-valued logic almost straightforward.

Łukasiewicz's basic idea was to supplement two-valued logic with a third truth value in such a way that the three-valued logic obtained would deviate least from ordinary logic. He did this by adopting the following truth table for implication:

Table 3.1 Truth values of implication $p \to q$ (p *implies q* or *if p then q*) in Łukasiewicz's three-valued logic

$\to$	0	$\frac{1}{2}$	1
0	1	1	1
$\frac{1}{2}$	$\frac{1}{2}$	1	1
1	0	$\frac{1}{2}$	1

and by assuming that the following formulae, which are in fact tautologies in two-valued logic, define, respectively, negation, disjunction, conjunction, and equivalence[2]:

$$\neg p \overset{def}{=} p \to 0 \tag{3.1}$$

(*not p* means the same as *p implies falsehood*)

$$\text{disjunction}: p \vee q \overset{def}{=} (p \to q) \to q \tag{3.2}$$

(*p or q* means the same as (*p implies q*) *implies q*)

$$\text{conjunction}: p \wedge q \overset{def}{=} \neg(\neg p \vee \neg q) \tag{3.3}$$

(*p and q* means the same as *not* (*not p or not q*))

$$\text{equivalence}: p \equiv q \overset{def}{=} (p \to q) \wedge (q \to p) \tag{3.4}$$

(*p and q are equivalent* means the same as *p implies q and q implies p*)

It can easily be checked that these definitions together with Table 3.1 yield the following truth tables for negation, disjunction, conjunction, and equivalence in the Łukasiewicz three-valued logic:

Table 3.2 Truth values of negation $\neg p$ (*not p*)

p	$\neg p$
0	1
$\frac{1}{2}$	$\frac{1}{2}$
1	0

---

[2]Łukasiewicz did not consider three-valued equivalence in [4]. In 1922 he did so for the first time in the realm of many-valued logics [8] but the formula (3.4) is a standard one in two-valued logic and it was often used in many of Łukasiewicz's papers published both before and after 1920.

Table 3.3 Truth values of disjunction $p \vee q$ (p or q)

$\vee$	0	$\frac{1}{2}$	1
0	0	$\frac{1}{2}$	1
$\frac{1}{2}$	$\frac{1}{2}$	$\frac{1}{2}$	1
1	1	1	1

Table 3.4 Truth values of conjunction $p \wedge q$ (p and q)

$\wedge$	0	$\frac{1}{2}$	1
0	0	0	0
$\frac{1}{2}$	0	$\frac{1}{2}$	$\frac{1}{2}$
1	0	$\frac{1}{2}$	1

Table 3.5 Truth values of equivalence $p \equiv q$ (p is equivalent to q or p if and only if q)

$\equiv$	0	$\frac{1}{2}$	1
0	1	$\frac{1}{2}$	0
$\frac{1}{2}$	$\frac{1}{2}$	1	$\frac{1}{2}$
1	0	$\frac{1}{2}$	1

When the first break-through was made, further generalization to n-valued logics and infinite-valued logics was not so difficult and Łukasiewicz actually did this soon afterwards [8, 9]. Of course except for the case of n-valued logics with n being a relatively small number, truth values of logical connectives cannot be presented in the form of tables. Fortunately, Łukasiewicz found algebraic expressions, the same for all systems of many-valued logics (both finite and infinite-valued), which yield truth values of compound propositions as functions of their constituents, i.e. all Łukasiewicz's many-valued logics are truth-functional. The basic logical functor for Łukasiewicz was implication and the following expression enables the truth value of implication $\tau(p \rightarrow q)$ to be calculated when the truth values $\tau(p)$ and $\tau(q)$ of the antecedent p and the consequent q are known [8, 9]:

$$\tau(p \rightarrow q) = \min[1 - \tau(p) + \tau(q), 1]. \tag{3.5}$$

It can be easily checked that this formula applied to definitions (3.1)–(3.4) yields the following formulae for truth values of the remaining logical connectives:

$$\tau(\neg p) = 1 - \tau(p) \tag{3.6}$$

$$\tau(p \vee q) = \max[\tau(p), \tau(q)] \tag{3.7}$$

$$\tau(p \wedge q) = \min[\tau(p), \tau(q)] \tag{3.8}$$

$$\tau(p \equiv q) = 1 - |\tau(p) - \tau(q)|. \tag{3.9}$$

Łukasiewicz assumed that the set of truth values of an n-valued logic consists of all fractions of the form $\frac{k}{n-1}$ with $0 \leq k \leq n-1$. It is straightforward to check that the formulae (3.6)–(3.9) yield Tables 3.1, 3.2, 3.3, 3.4 and 3.5 for $n = 3$.

Łukasiewicz mentioned in [8] that "...*0 is interpreted as falsehood, 1 as truth, and the other numbers in the interval 0–1 as the degrees of probability corresponding to various possibilities...*". Therefore, it is clear that at least in the year 1922 he still maintained his idea, expressed for the first time in the year 1913 [3], of interpreting non-classical truth values as degrees of probability.

In contrast to Jan Łukasiewicz, the second founding father of the modern theory of many-valued logics, Emil Leon Post, does not seem to be very much concerned with the interpretation of non-classical truth values. His investigations were not so much founded on philosophical considerations but were rather of a formal algebraic nature. Loosely speaking we can say that he studied algebraic aspects of n-valued logics without bothering to express them linguistically and in this respect his papers [5, 10] are closer in their style to modern treatises on many-valued logics (see, e.g. [11]) than contemporary papers by Łukasiewicz.

Post based his n-valued propositional calculi on the linearly ordered set of truth values $\{t_1, t_2, ...t_n\}$ where the extreme elements express "*full truth*" and "*full falsehood*" and he followed Whitehead and Russel's *Principia Mathematica* [12] in choosing negation and disjunction as basic connectives. However, although his disjunction was the same as that of Łukasiewicz, i.e. its truth value was the greater of the truth values of its constituents, Post's basic negation, which could be called *cyclic* was quite different (Table 3.6):

Table 3.6 Truth values of Post's "cyclic" negation

p	$\neg p$
t_1	t_2
t_2	t_3
$\vdots$	$\vdots$
t_n	t_1

With negation defined in this way,[3] if other connectives (conjunction, implication, and equivalence) are defined with the aid of tautologies taken from two-valued logic, then they exhibit rather unexpected and counterintuitive features. In spite of this fact, in modern times Post logics find their application in the study of electronic networks and in Computer Science [13, 14]. It should also be mentioned that due to this particular negation, the n-valued propositional calculi of Post, in contrast to those of Łukasiewicz, are functionally complete: any conceivable connective can be defined by basic connectives of negation and disjunction. Of course no intuitive interpretation can be given to the vast majority of connectives obtained in such a way.

---

[3]Besides this "cyclic" negation Post also considered the other negation, identical to that of Łukasiewicz. However, it seems that he treated "cyclic" negation as more important.

After the first breakthrough made by Łukasiewicz and Post many other systems of three-valued and n-valued propositional calculi were proposed. The three-valued calculi of Kleene [15, 16], Bochvar [17], and Finn [18] should be mentioned here, which were motivated by epistemological considerations concerning lack of meaning of some statements, and also a similarly motivated group of papers dealing with so-called nonsense-logics [19–22], attempts at describing intuitionistic propositional calculus in terms of many-valued logics [23, 24], or papers motivated by considerations concerning the peculiarities of quantum physics [25–32]. The mathematically experienced reader can find a detailed survey of most of the above-mentioned three-valued logics and some n-valued logics [33, 34] in Chaps. 3 and 4 of a book [11] by Bolc and Borowik. Some examples of many-valued logics motivated by physical considerations are described in Chap. 8 of Jammer's book [35].

It should be mentioned that Łukasiewicz's many-valued logic, endowed with negation (3.1), (3.6), disjunction (3.2), (3.7), and conjunction (3.3), (3.8) was criticized by Gonseth [36] in 1938 since it satisfies neither the law of the excluded middle

$$\tau(p \vee \neg p) = 1 \tag{3.10}$$

(it is true that p or not p)

nor the law of contradiction

$$\tau(p \wedge \neg p) = 0 \tag{3.11}$$

(it is false that p and not p).

In fact, neither of these formulae is satisfied for $\tau(p) \neq 0, 1$, as, for example, they both assume truth value $\frac{1}{2}$ for $\tau(p) = \frac{1}{2}$. Most probably Gonseth did not know Polish so he could not have read the paper [37] already published by Zawirski in 1934,[4] in which Zawirski noticed that if we replace the right-hand side of the formula (3.2) by which Łukasiewicz defined disjunction in [4] by the other (in two-valued logic equivalent) expression: $\neg p \rightarrow q$, then the disjunction obtained in this way:

$$p \sqcup q = \neg p \rightarrow q$$
$$\tau(p \sqcup q) = \min[\tau(p) + \tau(q), 1] \tag{3.12}$$

and the conjunction adjoint to it by de Morgan's Law:

$$p \sqcap q = \neg(\neg p \sqcup \neg q)$$
$$\tau(p \sqcap q) = \max[\tau(p) + \tau(q) - 1, 0] \tag{3.13}$$

satisfy both the law of the excluded middle:

---

[4]The same idea was published in English by Orrin Frink Jr. in 1938 in [38].

$$\tau(p \sqcup \neg p) = \min[\tau(p) + 1 - \tau(p), 1] = 1 \qquad (3.14)$$

and the law of contradiction:

$$\tau(p \sqcap \neg p) = \max[\tau(p) + 1 - \tau(p) - 1, 0] = 0. \qquad (3.15)$$

Therefore, Gonseth's critique cannot be applied to Łukasiewicz's many-valued logic endowed with his original implication (3.5), negation (3.1), (3.6), and the Zawirski-Frink disjunction (3.12) and conjunction (3.13). As we shall see in what follows, this set of connectives also seems to be better suited to the description of the behaviour of quantum physical systems than the set of connectives originally defined and studied by Łukasiewicz.

References

1. Łukasiewicz, J. "Philosophische Bemerkungen zu mehrwertigen Systemen des Aussagenkalküls", *Comptes rendus des séances de la Société des Sciences et des Lettres de Varsovie, Cl. III*, **23** (1930) 51–77; reprinted as "Philosophical remarks on many-valued systems of propositional logic" in [2], pp. 153–178.
2. Łukasiewicz, J. *Selected Works*, ed. by L. Borkowski (North-Holland, Amsterdam, and PWN—Polish Scientific Publishers, Warszawa, 1970).
3. Łukasiewicz, J. *Die logischen Grundlagen der Wahrscheinlichkeitsrechnung* (Acad. der Wiss. Kraków, 1913); reprinted as "Logical foundations of probability theory" in [2], pp. 16–63.
4. Łukasiewicz, J. "O logice trójwartościowej", *Ruch Filozoficzny*, **5** (1920) 170–171; reprinted as: "On three-valued logic" in [2], pp. 87–88.
5. Post, E. "Introduction to a general theory of elementary propositions", *Bulletin of the American Mathematical Society*, **26** (1920) 437.
6. Łukasiewicz, J. *An Address Delivered as a Rector of the Warsaw University at the Inauguration of the Academic Year 1922/1933*; reprinted as: "On determinism" in [2], pp. 110–128.
7. Łukasiewicz, J. *Farewell Lecture by Professor Jan Łukasiewicz Delivered in the Warsaw University Lecture Hall on March 7, 1918*; reprinted in [2], pp. 84–86.
8. Łukasiewicz, J. *Lecture delivered at the 232nd Meeting of the Polish Philosophical Society in Lwów on October 14, 1922*, published in *Ruch Filozoficzny*, **7** (1923) 92–93 (in Polish); reprinted as: "A numerical interpretation of the theory of propositions" in [2], pp. 129–130.
9. Łukasiewicz, J. and A. Tarski, "Untersuchungen über den Aussagenkalkül", *Comptes rendus des séances de la Société des Sciences et des Lettres de Varsovie, Cl. III*, **23** (1930) 39–50; reprinted as: "Investigations into the sentential calculus" in [2], pp. 131–152.
10. Post, E. "Introduction to a general theory of elementary propositions", *American Journal of Mathematics*, **43** (1921) 163–185.
11. Bolc, L. and P. Borowik, *Many-Valued Logics* (Springer-Verlag, Berlin, 1992).
12. Whitehead, A. N. and Russel, B. *Principia Mathematica* (Cambridge University Press, Cambridge, 1910).
13. Epstein, G., G. Frieder, and D. C. Rine, "The development of multiple-valued logic as related to Computer Science", *Computer*, **7** (1974) 20–32.
14. Rine, D. C. (ed), *Computer Science and Multiple-Valued Logic. Theory and Applications* (North-Holland, Amsterdam, 1977).
15. Kleene, S. C. "On a notation for ordinal numbers", *The Journal of Symbolic Logic*, **3** (1938) 150–155.
16. Kleene, S. C. *Introduction to Metamathematics* (North-Holland, Amsterdam, 1952).

17. Bochvar, D. A. "On a three-valued logical calculus and its application to the analysis of contradictions", *Matematičeskij Sbornik*, **4** (1939) 287–308 (in Russian).
18. Finn, V. K. *An Axiomatization of some Propositional Calculi and their Algebras* (Vsiechsojuznyi Institut Naučeskoi Informacii Akademii Nauk SSSR, Moskva, 1972) (in Russian).
19. Hallden, S. "The logic of nonsense", *Uppsala Universitets Arsskrift*, **9** (1949) 132.
20. Åqvist, L. "Reflections on the logic of nonsense", *Theoria*, **28** (1962) 138–158.
21. Segerberg, K. "A contribution to nonsense logic", *Theoria*, **31** (1965) 199–217.
22. Piróg-Rzepecka, K. *Systemy Nonsense-Logics* (PWN—Polish Scientific Publishers, Warszawa, 1977) (in Polish).
23. Heyting, A. *Intuitionism. An Introduction* (North-Holland, Amsterdam, 1966).
24. Jaśkowski, S. "Recherches sur le système de la logique intuitioniste", in: Actes du Congrès International de Philosophie Scientifique. Part 6, Philosophie des mathématiques. Paris, 1936. *Actualités scientifiques et industrielles*, **393** (1936) 58–61.
25. Zawirski, Z. "Jan Łukasiewicz 3-valued logic. On the logic of L. E. J. Brouwer. Attempts at applications of many-valued logic to contemporary natural science", *Sprawozdania Poznańskiego Towarzystwa Przyjaciół Nauk*, **2–4** (1931) 1–8 (in Polish).
26. Zawirski, Z. "Les logiques nouvelles et le champ de leur application", *Revue de Métaphisique et de Morale*, **39** (1932) 503–519.
27. Février, P. "Les relations d'incertitude de Heisenberg et la logique", *Comptes Rendus Acad. Sci. Paris*, **204** (1937) 481–483.
28. Destouches-Février, P. "Logiques et theories physiques", *Congrès International de Philosophie des Sciences, Paris 1949* (Herman, Paris, 1951) pp. 45–54.
29. Reichenbach, H. *Philosophic Foundations of Quantum Mechanics* (University of California Press, Berkeley, 1944).
30. Reichenbach, H. "The principle of anomaly in quantum mechanics", *Dialectica*, **2** (1948) 337–350.
31. Reichenbach, H. "Über die erkenntnistheoretische Problemlage und den Gebrauch einer dreiwertigen Logik in der Quantenmechanik", *Zeitschrift für Naturforschung*, **6a** (1951) 569–575.
32. Reichenbach, H. "Les fondements logiques de la mécanique des quanta", *Annales de l'Institut Henri Poincaré*, **13** (1952–1953) 109–158.
33. Słupecki, J. "Der volle dreiwertige Aussagenkalkül", *Comptes Rendus des Séances de la Société des Sciences et des Lettres de Varsovie, Classe III*, **29** (1936) 9–11.
34. Sobociński, B. "Axiomatization of certain many-valued systems of the theory of deduction", *Roczniki prac naukowych zrzeszenia asystentów Uniwersytetu Józefa Piłsudskiego w Warszawie*, **1** (1936) 399–419.
35. Jammer, M. *The Philosophy of Quantum Mechanics* (Wiley-Interscience, New York, 1974).
36. Gonseth, F. *Les entretiens de Zürich sur les fondements et la méthode des sciences mathematiques 6–9 décembre 1938* (Zürich, 1941).
37. Zawirski, Z. "Relationship between many-valued logic and the calculus of probability", *Prace Komisji Filozoficznej Poznańskiego Towarzystwa Przyjaciół Nauk*, **4** (1934) 155–240 (in Polish).
38. Frink, O. "New algebras of logic", *American Mathematics Monthly*, **45** (1938) 210–219.

Chapter 4
Fuzzy Sets and Many-Valued Logics

4.1 Rudiments of the Fuzzy Set Theory

Classical, two-valued logic is a basis of traditional mathematics and, in particular, of the traditional set theory. Although well-elaborated systems of axioms for the classical set theory do exist, for all practical purposes it is enough to distinguish a set that we are interested in by a predicate which, according to two-valued logic, enable all the objects under consideration to be unambiguously divided into two disjoint classes: objects that belong to a set and objects that do not belong to a set and form its complement. For example, let U be a set consisting of *speakers at the Conference on Foundations of Quantum Mechanics*. This predicate is precise enough to define this set as soon as the Conference is finished. All propositions of the form: *x belongs to the set U* where x denotes a name of an individual person are, as soon as the Conference is finished, either true or false, i.e. they belong to the domain of classical two-valued logic.

However, although every traditional set is defined by a "sharp" predicate, not every predicate is good enough to define a traditional set in an unambiguous way. Let us try to distinguish a subset A of the above-mentioned set of speakers U consisting of *speakers whose talks were interesting*. Even if we choose only one umpire in order not to deal with various opinions we are likely to get, besides "sharp" judgements of the form: *"the talk of Dr. X was not interesting"*, *"the talk of Dr. Y was interesting"* also a lot of statements of the form: *"the talk of ... was ...a little bit/only partially/not so much/quite/in most of its parts/almost... interesting"*. Therefore, we see that besides the speakers who, like Dr. X, surely do not belong to the set A and who, like Dr. Y, surely belong to it, both membership and non-membership of other speakers to the set A is doubtful. However, it would also not be good to group all these other speakers into one category since from the various judgements of our umpire we infer that different talks were interesting to him to a different extent. The best solution would be to evaluate numerically the degrees to which the talks of the various speakers were interesting and to say that the "degrees of membership" of the various speakers to the set A are proportional to these numbers.

© The Author(s) 2015
J. Pykacz, *Quantum Physics, Fuzzy Sets and Logic*,
SpringerBriefs in Physics, DOI 10.1007/978-3-319-19384-7_4

This is exactly the idea of a fuzzy set: If A is a fuzzy subset of the universe of discourse U (in our case the set U consists of all the speakers), then some elements of U surely belong to A, some surely do not belong to it, but all the intermediate cases of "partial membership" are also allowed. Moreover, membership is "graded": according to the original idea of Zadeh [1], who is generally recognized as a founding father of the fuzzy set theory,[1] membership of an element x to a fuzzy set A, denoted $\mu_A(x)$ or simply $A(x)$, can vary from 0 (full non-membership) to 1 (full membership) i.e., it can assume all values in the interval [0, 1]. Therefore, a *membership function* $\mu_A : x \mapsto \mu_A(x) \in [0, 1]$ completely characterizes the fuzzy set A and it is an obvious generalization of a characteristic function $\chi_A(x)$ of a traditional set:

$$\chi_A(x) = \begin{cases} 0 \text{ for } x \notin A \\ 1 \text{ for } x \in A \end{cases} \tag{4.1}$$

Fuzzy subsets of a plane can be easily visualized as areas which, contrary to traditional sets (usually called *crisp sets* in the fuzzy set theory), have no sharp boundaries and vanish gradually. They are *smeared*, *blurred* or simply *fuzzy*.

Our everyday language provides us with numerous examples of "non-sharp" predicates which can define only non-crisp sets, e.g. *young* (man), *ripe* (apple), *old* (painting), *fast* (car), *famous* (artist), etc. In all these cases we can easily distinguish elements which certainly belong to a set of objects defined by a given predicate, elements which surely do not belong to it, and elements whose membership is more or less doubtful. I would, in fact, venture to say that in everyday communication "sharp" predicates which define crisp sets are the exception rather than the rule. Of course, in some cases it is possible to draw a borderline in a more or less arbitrary way to recover sharp discrimination between members and non-members of a set. For example we could state that a car x_1 which can go faster than 150 km/h belongs to the set of fast cars which, according to two-valued logic implies that a car x_2 which can go at the most at 149,999 km/h is, by the very definition, not fast, so it does not belong to the set of fast cars. However, we feel that the car x_2 "almost belongs" to the set of fast cars and should not be treated in the same way as a car x_3 which can go at the most at 50 km/h. It is more natural to state that the grade of membership of the car x_2 to the set of fast cars is very close to 1 while the grade of membership of the car x_3 to this set is close to 0. Thus, the idea of representing the collection of fast cars in the form of a fuzzy set is very appealing, although it should be mentioned that in general it is not at all obvious precisely what a membership function of a specific fuzzy set should look like.[2]

As soon as membership functions of fuzzy sets are established, these sets are characterized to the full extent and we can define on them all relations and operations

---

[1]It seems that Chang [2] and Klaua [3] elaborated similar ideas independently of [1] and published them even slightly before [1]. It is for historians of science to explain why their papers, contrary to [1] are almost neglected.

[2]This observation gave rise to the notion of *probabilistic fuzzy sets* introduced by Hirota [4], whose membership functions are themselves "fuzzy". Of course this procedure can be continued, but objects obtained in this way are less and less convenient to deal with.

known from traditional set theory.[3] This is much to be expected since classical sets are actually special cases of fuzzy sets: they are fuzzy sets whose membership functions assume only two values: 0 and 1, i.e., these membership functions are in fact characteristic functions (4.1), and because all set-theoretic relations and operations on classical sets can be expressed in terms of their characteristic functions.

Definitions of the basic relations and operations on fuzzy sets were put forward early on by Zadeh in his historic paper [1] and these are still the most frequently used in all contributions to and applications of the fuzzy set theory. We shall see in what follows that Zadeh's intuitive choice was so natural because these operations follow from the connectives of Łukasiewicz's many-valued logic in exactly the same way as operations on classical sets follow from the connectives of classical logic.

Zadeh's basic relations and operations are defined with the aid of membership functions as follows (we assume, as is usually done in the fuzzy set theory, that all considered fuzzy sets are in fact fuzzy subsets of a fixed universe of discourse U):

Equality of fuzzy sets: $A = B$ iff for all elements x in the universe U

$$\mu_A(x) = \mu_B(x). \tag{4.2}$$

Inclusion of fuzzy sets: $A \subseteq B$ iff for all elements x in the universe U

$$\mu_A(x) \leq \mu_B(x). \tag{4.3}$$

Complement (negation) of a fuzzy set: A' is a complement of A iff for all elements x in the universe U

$$\mu_{A'}(x) = 1 - \mu_A(x). \tag{4.4}$$

Union (sum) of fuzzy sets: $A \cup B$ is a union of A and B iff for all elements x in the universe U

$$\mu_{A \cup B}(x) = \max[\mu_A(x), \mu_B(x)]. \tag{4.5}$$

Intersection (product) of fuzzy sets: $A \cap B$ is an intersection of A and B iff for all elements x of the universe U

$$\mu_{A \cap B}(x) = \min[\mu_A(x), \mu_B(x)]. \tag{4.6}$$

The fuzzy set theory is by no means only a mathematical game. Although in the beginning it was treated with some reserve by traditionally oriented "crisp" mathematicians, it quickly found numerous practical applications which vary from earthquake forecasting, computer medical diagnoses, decision making and pattern recognition to the production of control systems for the underground and more efficient vacuum cleaners. Moreover, it seems to be a very natural tool for all "soft" sciences which deal with vagueness or imprecision caused either, as in

---

[3]There are also operations which can be defined on fuzzy sets that have no counterparts in traditional set theory, e.g. the operation of "sharpening" a set which makes it "less fuzzy".

meteorology, by an excess of data or, as in economics, sociology, psychology, etc., by the human factor. In fact, the "applicational" aspect of fuzzy sets is maybe even better known than their theoretical aspects which still seem to be undervalued by "crisp" mathematicians.

4.2 Fuzzy Sets and Infinite-Valued Łukasiewicz Logic

In order to explain why the obvious links between fuzzy sets and many-valued Łukasiewicz logics were not studied[4] during the whole of the first decade of the rapid development of the fuzzy set theory, one should take into account two possible reasons. On the one hand Lotfi Zadeh and his followers seemed to be interested mostly in applications of the newly established theory and were not so much occupied with clarification of its foundations. On the other hand "pure" mathematicians of that time did not pay much attention to a theory which was probably seen by them as too simple in comparison with the sophisticated problems emerging on the very frontiers of contemporary "crisp" mathematics. In fact, the relation of classical logic to classical set theory, in particular definitions of set-theoretic operations in terms of the connectives of classical logic, are taught at the beginning of the secondary school. They remain the same when classical logic is replaced by infinite-valued Łukasiewicz logic and classical sets are replaced by fuzzy sets, but this observation was published by Giles [5] 10 years after the successful launching of the idea of fuzzy sets by Zadeh in 1965.

Let us recall the relations between propositions and sets known from school and see how they work when classical logic and classical sets are replaced by Łukasiewicz infinite-valued logic and fuzzy sets.

The notion of a set is adopted at school as a primitive notion and it is tacitly assumed that we know a set when we know all its elements. Therefore, any set A can be described as a collection of objects whose names turn a propositional function "x *belongs to* A" ("$x \in A$") into a true proposition. Symbolically:

$$A = \{x : \tau("x \in A") = 1\} \tag{4.7}$$

Because of the equality in the bracket the set A defined by the formula (4.7) is unavoidably crisp even if we replace classical two-valued logic by infinite-valued logic. However, if we rewrite this formula in the following form (in two-valued logic equivalent):

$$A = \{x : \tau("x \in A") \neq 0\} \tag{4.8}$$

and replace the classical two-valued logic by an infinite-valued logic, then, since in the infinite-valued logic a truth value of a non-false proposition can assume, besides 1, any value between 0 and 1, the set A turns out to be a fuzzy set with a membership

---

[4]Except almost neglected papers [2, 3] mentioned in footnote 1 of this chapter.

function defined by the truth values of propositions of the form "x *belongs to A*":

$$\mu_A(x) = \tau(\text{"}x \in A\text{"}). \tag{4.9}$$

(the degree of membership of an object x to the fuzzy set A is equal to the truth value of the proposition "x *belongs to A*")

This equality allows well-known secondary school definitions of the complement, union, and intersection of classical sets to be used together with Łukasiewicz's formulas for truth values of negation (3.6), disjunction (3.7), and conjunction (3.8) to justify Zadeh's intuitive choice of basic operations on fuzzy sets (4.4), (4.5), and (4.6):

Membership function of a complement (negation) of a fuzzy set:

$$\mu_{A'}(x) = \tau(\text{"}x \in A'\text{"}) = \tau(\text{"}x \notin A\text{"}) = \tau(\neg\text{"}x \in A\text{"})$$
$$= 1 - \tau(\text{"}x \in A\text{"}) = 1 - \mu_A(x). \tag{4.10}$$

Membership function of a union (sum) of fuzzy sets:

$$\mu_{A \cup B}(x) = \tau(\text{"}x \in A\text{"} \vee \text{"}x \in B\text{"})$$
$$= \max[\tau(\text{"}x \in A\text{"}), \tau(\text{"}x \in B\text{"})]$$
$$= \max[\mu_A(x), \mu_B(x)]. \tag{4.11}$$

Membership function of an intersection (product) of fuzzy sets:

$$\mu_{A \cap B}(x) = \tau(\text{"}x \in A\text{"} \wedge \text{"}x \in B\text{"})$$
$$= \min[\tau(\text{"}x \in A\text{"}), \tau(\text{"}x \in B\text{"})]$$
$$= \min[\mu_A(x), \mu_B(x)]. \tag{4.12}$$

However, as was mentioned at the end of Chap. 2, the original Łukasiewicz disjunction and conjunction are not the only conceivable connectives of this type. Therefore, the Zawirski-Frink disjunction (3.12) and conjunction (3.13) placed inside (4.11) and (4.12) yield other operations of union and intersection of fuzzy sets, called *bold union* and *intersection* by Giles [5] who was the first to study them within the fuzzy set theory (other names: *Giles, truncated, bounded, arithmetic, Łukasiewicz* operations)[5]:

---

[5]It is clear that Giles [5] was not aware of Zawirski's 1934 paper [6] where these operations appeared for the first time (in the domain of a many-valued logic), and he was also, most probably, not aware of Frink's 1938 paper [7]. It also seems that these operations were rediscovered many times by various authors which explains the multiplicity of their names. Although these operations did not appear explicitly in any Łukasiewicz paper, the name *Łukasiewicz operations* seems to be the most popular nowadays and will be used throughout this paper.

$$\mu_{A \cup B}(x) = \tau(\text{``}x \in A\text{''} \sqcup \text{``}x \in B\text{''})$$
$$= \min[\tau(\text{``}x \in A\text{''}) + \tau(\text{``}x \in B\text{''}), 1]$$
$$= \min[\mu_A(x) + \mu_B(x), 1] \tag{4.13}$$

$$\mu_{A \cap B}(x) = \tau(\text{``}x \in A\text{''} \sqcap \text{``}x \in B\text{''})$$
$$= \max[\tau(\text{``}x \in A\text{''}) + \tau(\text{``}x \in B\text{''}) - 1, 0]$$
$$= \max[\mu_A(x) + \mu_B(x) - 1, 0] \tag{4.14}$$

It is obvious that other disjunction-like and conjunction-like connectives of infinite-valued logic[6] define in the same way other operations of fuzzy set union and intersection and vice versa. All operations on fuzzy sets interpretable as fuzzy set union and intersection[7] yield disjunction-like and conjunction-like connectives of the infinite-valued Łukasiewicz logic.

References

1. Zadeh, L. A. "Fuzzy sets", *Information and Control*, **8** (1965) 338–353.
2. Chang, C. C. "Infinite valued logics as a basis of set theory", *Proceedings of the 1964 International Congress in Logic, Methodology, and Philosophy of Science* (North-Holland, Amsterdam, 1965) pp. 93–100.
3. Klaua, D. "Über einen Einsatz zur mehrwertigen Mengenlehre", *Monatsb. Deutsch. Acad. Wiss. Berlin*, **7** (1965) 859–867.
4. Hirota, M. "Concepts of probabilistic sets", *Proceedings of IEEE Conference on Decision and Control* (1977) 1361–1366; *Fuzzy Sets and Systems*, **5** (1981) 31–46.
5. Giles, R. "Łukasiewicz logic and fuzzy set theory", *International Journal of Man-Machine Studies*, **67** (1976) 313–327.
6. Zawirski, Z. "Jan Łukasiewicz 3-valued logic. On the logic of L. E. J. Brouwer. Attempts at applications of many-valued logic to contemporary natural science", *Sprawozdania Poznańskiego Towarzystwa Przyjaciół Nauk*, **2–4** (1931) 1–8 (in Polish).
7. Frink, O. "New algebras of logic", *American Mathematics Monthly*, **45** (1938) 210–219.
8. Frank, M. J. "On the simultaneous associativity of $F(x, y)$ and $x + y - F(x, y)$", *Aequationes Mathematicae*, **19** (1979) 194–226.
9. Yager, R. R. "On general class of fuzzy connectives", *Fuzzy Sets and Systems*, **4** (1986) 235–242.

---

[6]There are $2^{(2^2)} = 16$ conceivable two-argument connectives in 2-valued logic, $3^{(3^2)} = 19.683$ two-argument connectives in 3-valued logic, $n^{(n^2)}$ two-argument connectives in n-valued logic and obviously infinity of two-argument connectives in infinite-valued logic. Of course, not all of these could, in a reasonable way, be interpreted as a disjunction or a conjunction. Some of them could be interpreted as an implication or equivalence, but the overwhelming majority surely has no two-valued counterparts.

[7]Even the whole families of such operations, parametrized by real numbers, have already been studied (see, e.g. [8, 9]).

Chapter 5
Many-Valued Logics in Quantum Mechanics

In the years 1925–1926 the development of quantum physics itself experienced a "quantum jump": Under the influential works of Heisenberg, Schrödinger, Born, Jordan and Dirac [1–6] physicists abandoned the so-called "older quantum theory", which was merely an amalgamate of ideas and models taken from classical physics with the addition of ad hoc "quantum conditions", and developed quantum mechanics as an internally consistent, although mathematically highly sophisticated theory which, at least in its non-relativistic part, persists without drastic changes to the present. However, some implications of the new theory were so bizarre that there were scientists who claimed that quantum theory could not be comprehended on the grounds of classical two-valued logic.

The first to express such claims was a Polish logician, Zygmunt Zawirski, who was looking for possible fields of application for Łukasiewicz many-valued logic. In the papers published in 1931 [7] and in 1932 [8] Zawirski argued that the equivalence of "complementary theories", e.g. wave and particle pictures in the description of micro-objects, is possible only on the basis of (at least) three-valued logic, since in two-valued logic a statement such as *"light is a wave* AND *light consists of particles"* is a statement which is a conjunction of two propositions which cannot simultaneously be true. Therefore, according to the laws of classical two-valued logic such conjunction is necessarily a false proposition.

According to Zawirski, complementarity, typical of quantum mechanics, can be comprehended only on the basis of (at least) three-valued logic when we ascribe to two mutually exclusive theories a third truth value interpreted as "possibility" or "equal probability". Indeed, if a truth value of two propositions p and q equals $\frac{1}{2}$, then according to formula (3.8)

$$\tau(p \wedge q) = min\left[\frac{1}{2}, \frac{1}{2}\right] = \frac{1}{2} \tag{5.1}$$

so the conjunction of two "possible" statements is again "possible".

© The Author(s) 2015

J. Pykacz, *Quantum Physics, Fuzzy Sets and Logic*,

SpringerBriefs in Physics, DOI 10.1007/978-3-319-19384-7_5

Zawirski's papers: [7] published in Polish in a local journal, [8] published in French in a journal relatively little read by physicists and [9] published again in Polish, received little attention.[1]

More fortunate in propagating his ideas was the American astrophysicist Fritz Zwicky. His paper [10], published in the *Physical Review*, in which he gave physical arguments against the law of the excluded middle and in favour of the "many-valuedness of scientific truth", was much more widely discussed (see [11], pp. 345–346).

The best-known attempts at basing quantum mechanics on three-valued logics were elaborated in the forties and early fifties of the XX century by Paulette Destouches-Février [12, 13] and Hans Reichenbach [14–17].

Destouches-Février was undoubtedly influenced by the epistemological papers of a group of contemporary philosophers, G. Bachelard, F. Gonseth, P. Hertz and L. Rougier, for whom logic was an empirical science which may be changed when new experimental results are obtained. Besides two ordinary truth values (*true* and *false*) ascribed to propositions which, when checked experimentally, yield sometimes true and sometimes false judgements, she introduced a third truth value *absolutely false*. This third truth value she ascribed to propositions which can never, by their very nature, be confirmed experimentally, for example: "*the energy E has value E_0*" when E_0 does not belong to the energy spectrum. She used two types of conjunction of propositions about quantum systems, depending upon whether the propositions were associated with commuting or non-commuting observables. In the first case her conjunction coincided with the conjunction that Łukasiewicz had used in his three-valued logic ((3.3), Table 3.4 in Chap. 3). In the second case she argued on the basis of Heisenberg uncertainty relations that a conjunction of propositions associated with non-commuting observables is always *absolutely false*.

The best-elaborated and most widely discussed attempt at explaining quantum phenomena on the basis of three-valued logic was published by Hans Reichenbach in his book *Philosophic Foundations of Quantum Mechanics* [14]. He distinguished between *phenomena* = microphysical events connected with macroscopic events by "rather short casual chains" and *interphenomena* = interpolations between phenomena without direct manifestation in the form of a macroscopic effect. To illustrate this division, let us consider a typical two-slit experiment which, according to Feynman's well-known words "has in it the heart of quantum mechanics" [18]: Emission of a quantum object by a source and its absorption on a screen are phenomena, while its "path" between the source and the screen, i.e. everything that adherents of the orthodox Copenhagen interpretation forbid even to consider, belongs to the domain of interphenomena. Reichenbach argued that if someone wants to go beyond the

---

[1] Even in Poland: When in 1991 in the Polish National Library in Warsaw I had in my hands a copy of Zawirski's paper [9] it turned out that the pages of a booklet were still not cut apart, i.e. most probably no one had read this copy during the whole 60 years! A tribute should be paid to Max Jammer, who mentions Zawirski's papers in his famous book [11] on the philosophy of quantum mechanics. The interested reader will find in Chap. 8 of this book a more detailed historical survey of the applications of many-valued logics in the foundations of quantum mechanics up to the early seventies of the XX century.

Copenhagen interpretation, i.e. if he wants to describe interphenomena as well as phenomena, believing that they are governed by the same laws of nature, then application of two-valued logic inevitably leads to causal anomalies which vanish when bivalent logic is replaced by three-valued logic. According to Reichenbach the third truth value *indeterminate* should be treated ontologically and should not be confused with the macroscopic epistemological *unknown*.

Logical operations in Reichenbach's three-valued logic were defined independently of each other. He considered three types of negation, three types of implication, two types of equivalence, conjunction, and disjunction, out of which one type of negation (termed by Reichenbach diametrical), implication, and equivalence (called standard), conjunction, and disjunction, were identical with those of Łukasiewicz (3.1)–(3.4).

Reichenbach's ideas were pursued after his death by Putnam [19] and evoked much wider discussion than any other attempt at utilizing many-valued logic in the foundations of quantum mechanics, but critical voices prevailed (see [11], pp. 368–375). The same was true of the attitude of physicists to von Weizsäcker's "complementarity logic" [20] with complex truth values obtained directly from the mathematical formalism of Hilbert-space quantum mechanics.

As Max Jammer stated in his book [11], using the words quoted in the Introduction as the anti-motto to this book, von Weizsäcker's papers contained the last attempts to apply many-valued logics to foundations of quantum mechanics until the early seventies of the XX century. This does not mean that interest in applying non-classical logics to the foundations of quantum mechanics vanished completely. On the contrary, in the seventies and eighties more than thousand papers were published as well as nearly twenty books, and several international conferences on "quantum logics and related structures" were held (see the extensive Pavičić bibliography [21] published in 1992 that contains more than 1500 entries). Although no systematic bibliographical research was performed later, it seems that up to now this number could have even tripled. However, after von Weizsäcker's work the interest of researchers definitely shifted from three-valued or many-valued logics to the non-distributive but two-valued logic of the type proposed by Birkhoff and von Neumann as early as 1936 [22].

In the author's opinion the attempts to found quantum mechanics on any version of three-valued logics were bound to fail also because three-valued logics are not "rich enough in truth values" and, therefore, are not "flexible" enough. One should not expect to be able to describe by these means the whole variety of quantum phenomena, especially when truth values are supposed to be connected with the numerical results of experiments, if there is only one truth value (besides the classical 0 and 1) available. On the other hand, von Weizsäcker's complex-valued "complementarity logic" seems to be too different from anything that could be intuitively accepted as "logic". Nevertheless, the possibility of describing quantum phenomena by means of many-valued logic still remains. The aim of the next two chapters is to show that the "orthodox" Birkhoff–von Neumann quantum logic can be treated equivalently as a special kind of infinite-valued Łukasiewicz logic.

References

1. Heisenberg, W. "Über quantentheotetische Umdeutung kinematischer und mechanischer Beziehungen", *Zeitschrift für Physik.* **33** 879–893.
2. Schrödinger, E. "Quantisierung als Eigenwertproblem", *Annalen der Physik*, **79** (1926) 361–376; 499–507; **80** (1926) 437–490; **81** (1926) 109–139.
3. Born, M., and P. Jordan, "Zur Quantenmechanik", *Zeitschrift für Physik*, **34** (1925) 858–888.
4. Born, M., W. Heisenberg, and P. Jordan, "Zur Quantenmechanik II", *Zeitschrift für Physik*, **35** (1926) 557–615.
5. Born, M. "Zur Quantenmechanik der Stossvorgänge", *Zeitschrift für Physik*, **37** (1926) 863–867.
6. Dirac, P. A. M. "The fundamental equations of quantum mechanics", *Proc. Royal Society of London (A)*, **109** (1925) 642–653.
7. Zawirski, Z. "Jan Łukasiewicz 3-valued logic. On the logic of L. E. J. Brouwer. Attempts at applications of many-valued logic to contemporary natural science", *Sprawozdania Poznańskiego Towarzystwa Przyjaciół Nauk*, **2–4** (1931) 1–8 (in Polish).
8. Zawirski, Z. "Les logiques nouvelles et le champ de leur application", *Revue de Métaphisique et de Morale*, **39** (1932) 503–519.
9. Zawirski, Z. "Relationship between many-valued logic and the calculus of probability", *Prace Komisji Filozoficznej Poznańskiego Towarzystwa Przyjaciół Nauk*, **4** (1934) 155–240 (in Polish).
10. Zwicky, F. "On a new type of reasoning and some of its possible consequences", *Physical Review*, **43** (1933) 1031–1033.
11. Jammer, M. *The Philosophy of Quantum Mechanics* (Wiley-Interscience, New York, 1974).
12. Février, P. "Les relations d'incertitude de Heisenberg et la logique", *Comptes rendus Acad. Sci. Paris*, **204** (1937) 481–483.
13. Destouches-Février, P. "Logiques et theories physiques", *Congrès International de Philosophie des Sciences, Paris 1949* (Herman, Paris, 1951) pp. 45–54.
14. Reichenbach, H. *Philosophic Foundations of Quantum Mechanics* (University of California Press, Berkeley, 1944).
15. Reichenbach, H. "The principle of anomaly in quantum mechanics", *Dialectica*, **2** (1948) 337–350.
16. Reichenbach, H. "Über die erkenntnistheoretische Problemlage und den Gebrauch einer dreiwertigen Logik in der Quantenmechanik", *Zeitschrift für Naturforschung*, **6a** (1951) 569–575.
17. Reichenbach, H. "Les fondements logiques de la mécanique des quanta", *Annales de l'Institut Henri Poincaré*, **13** (1952–1953) 109–158.
18. Feynman, R. P., R. B. Leighton, and M. Sands, *The Feynmann Lectures on Physics* (Addison-Wesley, Reading, MA, 1965) Vol III, p. 1–1.
19. Putnam, H. "Three-valued logic", *Philosophical Studies*, **8** (1957) 73–80.
20. von Weizsäcker, C. F. "Die Quantentheorie der einfachen Alternative", *Zeitschrift für Naturforschung*, **13a** (1958) 245–253.
21. Pavičić, M. "Bibliography on quantum logics and related structures", *International Journal of Theoretical Physics*, **31** (1992) 373–461.
22. Birkhoff, G. and J. von Neumann, "The logic of quantum mechanics", *Annals of Mathematics*, **37** (1936) 823–843.

Chapter 6
Birkhoff-von Neumann Quantum Logic

According to the historic Birkhoff and von Neumann paper [1] a family of experimental propositions pertaining to a quantum system should possess an algebraic structure characteristic of the family of all linear subspaces of a (finite-dimensional) Hilbert space, i.e. it should be an orthocomplemented modular lattice. Since in the case of infinite-dimensional Hilbert spaces lattices of their closed linear subspaces are not modular, the requirement of modularity was soon replaced by Husimi [2] by a more general orthomodularity condition, satisfied also in this case. In some versions of the logico-algebraic approach to foundations of quantum mechanics the algebraic model of a family of experimental propositions about a physical system is assumed to be a slightly more general orthomodular partially ordered set instead of an orthomodular lattice. This assumption is adopted also throughout this book, although it should be mentioned that by making it we move a little further from the very archetype, namely an orthomodular lattice of all closed linear subspaces of a (possibly infinite-dimensional) separable Hilbert space or, equivalently, a lattice of orthogonal projectors onto these subspaces.

We shall quote now the full definition of the basic algebraic structure we shall deal with. This object, believed to be an algebraic representation of a set of experimental propositions pertaining to a physical system is traditionally called *quantum logic*. In our opinion this name is somewhat unfortunate since it identifies the "logic of experimentally verifiable propositions" with its algebraic representation. Moreover, since Boolean algebras, which are algebraic representations of sets of propositions about classical physical systems, also belong to this class, even the very word "quantum" as understood in "quantum logic" is misleading. Therefore, the name "orthomodular algebra" coined by Burmeister and Mączyński [3] seems to be far preferable, also because it corresponds very well to other "algebras" (Boolean algebras, orthoalgebras, effect algebras, generalized MV algebras, etc.) studied within the modern theory of quantum structures. Unfortunately, this term is not yet sufficiently popular, so throughout this book we shall use the more popular traditional name.

© The Author(s) 2015
J. Pykacz, *Quantum Physics, Fuzzy Sets and Logic*,
SpringerBriefs in Physics, DOI 10.1007/978-3-319-19384-7_6

6.1 The Traditional Algebraic Model

Following the well-established tradition (cf. for example, [4, 5]), by a *quantum logic* we mean an orthocomplemented σ-orthocomplete orthomodular poset, i.e. a partially ordered set L which contains the smallest element O and the greatest element I, in which the orthocomplementation map $\perp: L \to L$ satisfying the conditions (a)–(c) exists:

(a) $(a^{\perp})^{\perp} = a$.
(b) If $a \leq b$, then $b^{\perp} \leq a^{\perp}$.
(c) The greatest lower bound (*meet*) $a \wedge a^{\perp}$ and the least upper bound (*join*) $a \vee a^{\perp}$ with respect to the given partial order exist in L and $a \wedge a^{\perp} = O, a \vee a^{\perp} = I$.

Moreover, the σ-orthocompleteness condition holds:

(d) If $a_i \leq a_j^{\perp}$ for $i \neq j$ (such elements are called *orthogonal* and are usually denoted $a_i \perp a_j$), then the join $\vee_i a_i$ exists in L,

and so does the orthomodular identity:

(e) If $a \leq b$, then $b = a \vee (a^{\perp} \wedge b) = a \vee (a \vee b^{\perp})^{\perp}$.

Elements $a, b \in L$ are called *compatible* iff there exist in L pairwise orthogonal elements a_1, b_1, c such that $a = a_1 \vee c$ and $b = b_1 \vee c$.

Probability measure on a quantum logic L is a mapping $s : L \to [0, 1]$ such that

(i) $s(I) = 1$,
(ii) $s(\vee_i a_i) = \Sigma_i s(a_i)$ for any sequence of pairwise orthogonal elements of L.

If elements of a quantum logic L represent experimentally verifiable propositions about a physical system, then probability measures defined on L represent states of a physical system and, therefore, are often themselves called *states* on L. According to the standard interpretation a number $s(a) \in [0, 1]$ is interpreted as a probability of obtaining the result "yes" in an experiment designed to check a proposition represented by a when a physical system is in a state represented by s. However, as we shall see in what follows, this number can also be interpreted as a truth value of a many-valued proposition about results of not-yet-performed experiments: *"an experiment designed to check a property represented by a will show that a system in a state represented by s actually has this property"* or simply a truth value of a many-valued proposition *"a physical system in a state represented by s has a property represented by a"*.

A set of probability measures (states) S on a quantum logic L is called *ordering* (*full, order determining*) iff $s(a) \leq s(b)$ for all $s \in S$ implies $a \leq b$. Let us note that the only way in which one can establish experimentally the partial order relation between various propositions is to conduct experiments on a system prepared in various states, which means that only quantum logics allowing for ordering sets of probability measures can be endowed with a physical interpretation. Therefore, throughout the rest of the book we shall consider only quantum logics with ordering sets of probability measures.

6.2 Mączyński's Functional Model

In 1973, 37 years after the introduction of the very notion of quantum logic by Birkhoff and von Neumann [1], Mączyński published [6] a theorem which 20 years later turned out to be a milestone in building fuzzy set or, equivalently, many-valued models of B-vN quantum logics endowed with ordering sets of probability measures. This theorem, expressed by Mączyński in [7] in an equivalent form that is better suited for our purposes, appears as follows:

Theorem 1 *Let S be a non-empty set and let L be a set of mappings from S into* [0, 1] *that has the following three properties:*

(i) 0 *(the null function) belongs to* L
(ii) a ∈ L *implies* 1 − a ∈ L
(iii) *for any (finite or countable) sequence* $a_1, a_2, ..., a_i \in L$ *such that* $a_i + a_j \leq 1$ *for* $i \neq j$ *(such functions were called in [6] pairwise orthogonal), we have* $a_1 + a_2 + ... \in L$.

Then L is a quantum logic with respect to the natural partial order of real functions, with orthocomplementation $a^\perp = 1 - a$. *Every point* $u \in S$ *induces a probability measure* m_u *on* $(L, \leq, ^\perp)$, *where* $m_u(a) = a(u)$ *for all* $a \in L$, *and the family of measures* $\{m_u : u \in S\}$ *is ordering.*

Conversely, if $(L, \leq, ^\perp)$ *is a quantum logic with an ordering set S of probability measures, then each* $a \in L$ *induces a function* $\overline{a} : S \rightarrow [0, 1]$ *where* $\overline{a}(m) = m(a)$ *for all* $m \in S$. *The set of all such functions* $\overline{L} = \{\overline{a} : a \in L\}$ *has properties (i)–(iii) and* $(\overline{L}, \leq, ^\perp)$ *is isomorphic to* $(L, \leq, ^\perp)$.

Mączyński's functional representation theorem provides a very useful tool for studying quantum logics since it allows abstract algebraic and order-theoretic notions that appear in the original definition of a quantum logic to be expressed in a more convenient language of real functions. Moreover, since any [0, 1]-valued function can be thought of as a membership function of a fuzzy set, Mączyński's theorem allows any B-vN quantum logic endowed with an ordering set of probability measures to be represented in the form of a family of fuzzy sets, which further enables it to be treated as a kind of the infinite-valued Łukasiewicz logic. Indeed, if we think of the Mączyński functionals as membership functions of fuzzy sets, then condition (i) means that the empty set belongs to L, while condition (ii) means that if a fuzzy set belongs to L, then its standard fuzzy complement (4.4) also belongs to L. Only the third of Mączyński's conditions cannot be directly expressed in terms of fuzzy set notions. However, we shall show in the next section that this obstacle can be overcome at the expense of adding one more, very natural, condition.

6.3 The General Fuzzy Set Model

The formal similarity of some operations on fuzzy sets to order-theoretic operations on quantum logics yielded in the late eighties of the XX century the idea of
constructing fuzzy set models of B-vN quantum logics [8]. It was obvious from the
very beginning that orthocomplementation should be modelled by the standard fuzzy
set complementation (4.4) which, however, excludes the possibility of representing
meets and joins by Zadeh's unions (4.5) and intersections (4.6) [9]. Indeed, the formal fuzzy set counterparts of the excluded middle law and the law of contradiction
that are assumed in the point (c) of the definition of the B-vN quantum logic:

$$A \cap A' = \emptyset \qquad\qquad\qquad (6.1)$$

$$A \cup A' = \mathcal{U} \qquad\qquad\qquad (6.2)$$

are not satisfied by any genuine fuzzy (i.e., non-crisp) set, but they are satisfied if we
replace Zadeh's operations (4.5), (4.6) by Łukasiewicz's operations (4.13), (4.14).
Also the fact that the Zadeh operations are distributive does not allow for their use
in order to build fuzzy set models of generally non-distributive quantum logics.

The attempts at building fuzzy set models of quantum logics attempted by the
author since 1987 were in the beginning flawed by the necessity of using, together
with genuine fuzzy set operations (4.4), (4.13), and (4.14), pointwise algebraic sums
of membership functions which are alien to the fuzzy set theory since they may yield
outcomes bigger than 1. This drawback was overcome in 1994 [10] at the expense of a
slight modification of the third of the original Mączyński [6, 7] conditions described
in the previous section, and the addition of the natural requirement that an empty set
is the only fuzzy set that is (weakly) disjoint with itself.

Mączyński's functional representation theorem quoted in the previous section,
expressed in the language of fuzzy sets, and combined with the author's modification
yield the following theorem proved in [10]:

Theorem 2 *Any quantum logic L with an ordering set of probability measures S
can be isomorphically represented in the form of a family $\mathcal{L}(S)$ of fuzzy subsets of S
satisfying the following conditions:*

(a) $\mathcal{L}(S)$ contains the empty set $\emptyset$, i.e. such set that $\mu_\emptyset(s) = 0$ for all $s \in S$.
*(b) $\mathcal{L}(S)$ is closed with respect to the standard fuzzy set complementation (4.4), i.e.,
    if $A \in \mathcal{L}(S)$, then $A' \in \mathcal{L}(S)$.*
*(c) $\mathcal{L}(S)$ is closed with respect to the countable Łukasiewicz unions (4.13) of pair-
    wise weakly disjoint [11] sets, i.e., such sets that their Łukasiewicz's intersection
    is the empty set. In symbols: if $A_i \sqcap A_j = \emptyset$ for $i \neq j$, then $\sqcup_i A_i \in \mathcal{L}(S)$.*
*(d) The empty set $\emptyset$ is the only set in $\mathcal{L}(S)$ that is weakly disjoint with itself, i.e., for
    any $A \in \mathcal{L}(S)$, if $A \sqcap A = \emptyset$, then $A = \emptyset$.*

*Conversely, any family of fuzzy subsets of an arbitrary universe $\mathcal{U}$ satisfying con-
ditions (a)–(d) is a quantum logic partially ordered by the inclusion of fuzzy sets (4.3),*

with the fuzzy set complementation (4.4) as orthocomplementation, the orthogonality of the elements coinciding with their weak disjointness, and an ordering set of probability measures generated by points of the universe $\mathcal{U}$ according to the formula

$$s_x(A) = \mu_A(x) \ \text{for all} \ x \in \mathcal{U}. \tag{6.3}$$

Following the widespread custom we shall, when no confusion arises, identify fuzzy sets with their membership functions and we shall write $A, B, \ldots$ or $A(x), B(x), \ldots$ instead of $\mu_A, \mu_B, \ldots$ or $\mu_A(x), \mu_B(x), \ldots$.

Proof The proof is a modification of the original proof given in [10].

Because of Theorem 1 it is enough to prove that the Mączyński conditions (i)–(iii), when applied to membership functions of fuzzy subsets of S (or $\mathcal{U}$), are equivalent to conditions (a)–(d).

The equivalence of conditions (i) and (a), and conditions (ii) and (b) is obvious. The implication (iii) $\Rightarrow$ (c) is obvious as well, since in this case the membership function of the Łukasiewicz union in (c) equals the pointwise sum of membership functions of its constituents.

The fact that the condition (d) follows from (i)–(iii) was already proved by the author in [9] (Theorem 3.2). Indeed, $A \sqcap A = \emptyset$ means that $2A \leq 1$, so $A \leq 1/2 \leq 1 - A = A'$. Therefore, $A \wedge A' = A$ which, since by Theorem 1 any family of functions that satisfies (i)–(iii) is a quantum logic in which $A' = A^\perp$, means that $A = \emptyset$.

In order to finish the proof it is enough to show that conditions (a)–(d) imply (iii). First, let us notice that Giles' weak disjointness of fuzzy sets: $A_i \sqcap A_j = \emptyset$ is equivalent to Mączyński's pairwise orthogonality of their membership functions: $A_i + A_j \leq 1$. It is obvious, that for any sequence of pairwise weakly disjoint fuzzy sets, if $\sum_i A_i \leq 1$, then $\sum_i A_i = \sqcup_i A_i \in \mathcal{L}(S)$. Therefore, it is enough to show that for any sequence of such fuzzy sets conditions (a)–(d) make $\sum_i A_i > 1$ impossible. The proof of this fact will proceed by induction.

As has already been mentioned, for $n = 2$, $A_1 \sqcap A_2 = \emptyset$ is equivalent to $A_1 + A_2 \leq 1$ by the very definition of the Łukasiewicz product (4.14).

Let us assume that $\sum_{i=1}^{n} A_i \leq 1$ for any sequence of pairwise weakly disjoint sets of the length n. Let $\{A_i\}_{i=1}^{n+1}$ be any sequence of pairwise weakly disjoint sets of the length $n + 1$. By the induction hypothesis we can write down $n + 1$ inequalities:

$$
\begin{array}{l}
A_2 + A_3 + A_4 + \ldots + A_n + A_{n+1} \leq 1 \\
A_1 + A_3 + A_4 + \ldots + A_n + A_{n+1} \leq 1 \\
A_1 + A_2 + A_4 + \ldots + A_n + A_{n+1} \leq 1 \\
\qquad\qquad\qquad \vdots \\
A_1 + A_2 + A_3 + \ldots + A_{n-1} + A_{n+1} \leq 1 \\
A_1 + A_2 + A_3 + \ldots + A_{n-1} + A_n \leq 1
\end{array}
$$

After summing them up and dividing by n we obtain

$$\sum_{i=1}^{n+1} A_i \le \frac{n+1}{n}. \tag{6.4}$$

Let us denote $B_n = \sqcup_{i=1}^{n} A_i = \sum_{i=1}^{n} A_i$ and $B_{n+1} = \sqcup_{i=1}^{n+1} A_i = B_n \sqcup A_{n+1}$, and calculate $\left(B_{n+1}^{\perp} + A_{n+1}\right)(x)$. There are two possibilities:

1. If $\sum_{i=1}^{n+1} A_i(x) = B_n(x) + A_{n+1}(x) > 1$, then

$$\begin{aligned}
\left(B_{n+1}' + A_{n+1}\right)(x) &= \left[1 - (B_n \sqcup A_{n+1}) + A_{n+1}\right](x) \\
&= 1 - \min\left[B_n(x) + A_{n+1}(x), 1\right] + A_{n+1}(x) \\
&= 1 - 1 + A_{n+1}(x) = A_{n+1}(x) \le 1.
\end{aligned} \tag{6.5}$$

2. If $\sum_{i=1}^{n+1} A_i(x) = B_n(x) + A_{n+1}(x) \le 1$, then

$$\begin{aligned}
\left(B_{n+1}' + A_{n+1}\right)(x) &= 1 - \min\left[B_n(x) + A_{n+1}(x), 1\right] + A_{n+1}(x) \\
&= 1 - B_n(x) \le 1.
\end{aligned} \tag{6.6}$$

We see that in both cases $\left(B_{n+1}' + A_{n+1}\right)(x) \le 1$. This means that $B_{n+1}' \sqcap A_{n+1} = \emptyset$, so by the conditions (c) and (b) both $B_{n+1}' \sqcup A_{n+1}$, and $\left(B_{n+1}' \sqcup A_{n+1}\right)'$ belong to the family of fuzzy sets distinguished by (a)–(d). Let us calculate now $\left[\left(B_{n+1}' \sqcup A_{n+1}\right)' + B_n'\right](x)$, considering the same two possibilities as before and taking into account that in both cases

$$B_{n+1}' \sqcup A_{n+1} = B_{n+1}' + A_{n+1} \tag{6.7}$$

so we can use (6.5) or (6.6), respectively.

1. If $\sum_{i=1}^{n+1} A_i(x) > 1$, then

$$\begin{aligned}
\left[\left(B_{n+1}' \sqcup A_{n+1}\right)' + B_n'\right](x) &= A_{n+1}'(x) + B_n'(x) \\
&= 1 - A_{n+1}(x) + 1 - B_n(x) \\
&= 2 - \left[B_n(x) + A_{n+1}(x)\right] \\
&= 2 - \sum_{i=1}^{n+1} A_i(x) < 1.
\end{aligned} \tag{6.8}$$

2. If $\sum_{i=1}^{n+1} A_i(x) \le 1$, then

$$\left[\left(B_{n+1}' \sqcup A_{n+1}\right)' + B_n'\right](x) = B_n''(x) + B_n'(x) = B_n(x) + B_n'(x) = 1. \tag{6.9}$$

Therefore, again in both cases $\left[\left(B'_{n+1} \sqcup A_{n+1}\right)' + B'_n\right](x) \leq 1$, which means that $\left(B'_{n+1} \sqcup A_{n+1}\right)' \sqcup B'_n$ belongs to the distinguished family of fuzzy sets. However, combining (6.4) with (6.8) and (6.9), we obtain for any $n \geq 2$ and any x

$$\frac{1}{2} \leq 2 - \frac{n+1}{n} \leq \left[\left(B'_{n+1} \sqcup A_{n+1}\right)' \sqcup B'_n\right](x) \leq 1.$$

This means that for any x, $C(x) = \left[\left(B'_{n+1} \sqcup A_{n+1}\right)' \sqcup B'_n\right]'(x) \leq \frac{1}{2}$, so $C \sqcap C = \emptyset$ which, by (d) means that $C = \emptyset$. Now, if we assume that there exists x such that $\sum_{i=1}^{n+1} A_i(x) > 1$, we have, according to (6.7) and (6.5)

$$\left[\left(B'_{n+1} \sqcup A_{n+1}\right) \sqcap B_n\right](x) = \max\left[\left(B'_{n+1} + A_{n+1}\right)(x) + B_n(x) - 1, 0\right]$$
$$= \max\left[A_{n+1}(x) + B_n(x) - 1, 0\right]$$
$$= \max\left[\sum_{i=1}^{n+1} A_i(x) - 1, 0\right]$$
$$= \sum_{i=1}^{n+1} A_i(x) - 1 \neq 0,$$

so $C = \left(B'_{n+1} \sqcup A_{n+1}\right) \sqcap B_n \neq \emptyset$, contrary to (d).

Since for a countable sequence $\sum_i A_i$ is a pointwise limit of finite sums, we infer that for any sequence of pairwise weakly disjoint sets $\{A_i\}$ the assumption that there exists x such that $\sum_i A_i(x) > 1$ inevitably implies that the condition (d) cannot be satisfied, which finishes the proof. $\qquad\square$

Because of the second part of Theorem 2, any family $\mathcal{L}(\mathcal{U})$ of fuzzy subsets of an arbitrary universe $\mathcal{U}$ satisfying conditions (a)–(d) of Theorem 2, i.e., such that

(a) $\emptyset \in \mathcal{L}(\mathcal{U})$,
(b) if $A \in \mathcal{L}(\mathcal{U})$, then $A' \in \mathcal{L}(\mathcal{U})$,
(c) if $A_i \sqcap A_j = \emptyset$ for $i \neq j$, then $\sqcup_i A_i \in \mathcal{L}(\mathcal{U})$,
(d) if $A \sqcap A = \emptyset$, then $A = \emptyset$,

will be called a *quantum logic of fuzzy sets* or simply a *fuzzy quantum logic*. Of course the second part of Theorem 2 implies that any quantum logic of fuzzy sets $\mathcal{L}(\mathcal{U})$ has an ordering set of probability measures generated by points of the universe $\mathcal{U}$ according to the formula (6.3).

6.4 Two Pairs of Binary Operations

Since any fuzzy quantum logic is a σ-orthomodular poset with respect to the standard fuzzy set inclusion as partial order, it is endowed with two pairs of binary operations: the Łukasiewicz union $\sqcup$ and intersection $\sqcap$, by the aid of which it is defined, and the ordinary join $\vee$ and meet $\wedge$ with respect to the partial order $\subseteq$. Of course, since by Theorem 2 any quantum logic with an ordering set of probability measures can be isomorphically represented as a fuzzy quantum logic, one can also think of it as equipped with these two pairs of binary operations. Therefore, it is of the utmost importance to study the relations between these operations and also seek for their logical interpretation.

The theorems of this section were announced without proofs in the author's talk given at the 10th International Congress of Logic, Methodology, and Philosophy of Science (Firenze, August 1995) [12]. However, due to an unexpected delay, proofs were only published 5 years later in [13].

Theorem 3 *Let $\mathcal{L}(\mathcal{U})$ be a quantum logic of fuzzy subsets of a universe $\mathcal{U}$ and let $A, B \in \mathcal{L}(\mathcal{U})$. Then $A \sqcap B \in \mathcal{L}(\mathcal{U})$ iff $A \sqcup B \in \mathcal{L}(\mathcal{U})$, and in this case A and B are compatible, $A \sqcap B = A \wedge B$, and $A \sqcup B = A \vee B$.*

Before we proceed to the proof of Theorem 3, we shall prove the following lemma:

Lemma 1 *Let $\mathcal{L}(\mathcal{U})$ be a quantum logic of fuzzy subsets of a universe $\mathcal{U}$ and let $A, B \in \mathcal{L}(\mathcal{U})$. Then*

(a) If both $A \vee B \in \mathcal{L}(\mathcal{U})$ and $A \sqcup B \in \mathcal{L}(\mathcal{U})$, then $A \vee B = A \sqcup B$.
(b) If both $A \wedge B \in \mathcal{L}(\mathcal{U})$ and $A \sqcap B \in \mathcal{L}(\mathcal{U})$, then $A \wedge B = A \sqcap B$.

Proof We shall prove first that for any two fuzzy subsets A, B of the same universe $\mathcal{U}$

(i) $(A \sqcup B)(x) - (A \cup B)(x) \leq \frac{1}{2}$
    and
(ii) $(A \cap B)(x) - (A \sqcap B)(x) \leq \frac{1}{2}$

for any $x \in \mathcal{U}$, where $A \cup B$ and $A \cap B$ denote, respectively, Zadeh union (4.5) and Zadeh intersection (4.6) of fuzzy sets A and B.

(i) Let us calculate the difference

$$D_i(x) = (A \sqcup B)(x) - (A \cup B)(x)$$
$$= \min[A(x) + B(x), 1] - \max[A(x), B(x)]. \qquad (6.10)$$

There are four possibilities:

1. $A(x) \leq B(x)$ and $A(x) + B(x) \leq 1$, then $A(x) \leq \frac{1}{2}$ and $D_i(x) = A(x) + B(x) - B(x) \leq \frac{1}{2}$.
2. $A(x) \leq B(x)$ and $A(x) + B(x) \geq 1$, then $B(x) \geq \frac{1}{2}$ and $D_i(x) = 1 - B(x) \leq \frac{1}{2}$.
3. $A(x) \geq B(x)$ and $A(x) + B(x) \leq 1$, then $B(x) \leq \frac{1}{2}$ and $D_i(x) = A(x) + B(x) - A(x) \leq \frac{1}{2}$.
4. $A(x) \geq B(x)$ and $A(x) + B(x) \geq 1$, then $A(x) \geq \frac{1}{2}$ and $D_i(x) = 1 - A(x) \leq \frac{1}{2}$.

(ii) Let us calculate the difference

$$D_{ii}(x) = (A \cap B)(x) - (A \sqcap B)(x)$$
$$= \min[A(x), B(x)] - \max[A(x) + B(x) - 1, 0]. \qquad (6.11)$$

This difference takes the following values in the four cases already considered:

1. $D_{ii}(x) = A(x) - 0 = A(x) \le \frac{1}{2}$.
2. $D_{ii}(x) = A(x) - [A(x) + B(x) - 1] = 1 - B(x) \le \frac{1}{2}$.
3. $D_{ii}(x) = B(x) - 0 = B(x) \le \frac{1}{2}$.
4. $D_{ii}(x) = B(x) - [A(x) + B(x) - 1] = 1 - A(x) \le \frac{1}{2}$.

Proof of part (a).

Let us assume that both $A \vee B$ and $A \sqcup B$ belong to $\mathcal{L}(\mathcal{U})$. Since $(A \cup B)(x)$ is a pointwise supremum of membership functions $A(x)$ and $B(x)$, i.e., a supremum in the class of all possible functions that map the universe $\mathcal{U}$ into $[0, 1]$, while $(A \vee B)(x)$ is a supremum in the restricted class of membership functions of elements of $\mathcal{L}(\mathcal{U})$, the following inequalities hold for any x in $\mathcal{L}(\mathcal{U})$.

$$(A \cup B)(x) \le (A \vee B)(x) \le (A \sqcup B)(x) \qquad (6.12)$$

Lemma 3.1 of [9] states that if $E, F \in \mathcal{L}(\mathcal{U})$ and $E \subseteq F$, then $F - E \in \mathcal{L}(\mathcal{U})$, where $F - E$ is defined by the pointwise difference of respective membership functions:

$$(F - E)(x) = F(x) - E(x) \text{ for all } x \in \mathcal{L}(\mathcal{U}). \qquad (6.13)$$

Therefore, the fuzzy set $D = A \sqcup B - A \vee B$ belongs to $\mathcal{L}(\mathcal{U})$ and by (i) for any $x \in \mathcal{U}$

$$D(x) = (A \sqcup B)(x) - (A \vee B)(x) \le (A \sqcup B)(x) - (A \cup B)(x) \le \frac{1}{2}. \qquad (6.14)$$

Therefore, for any $x \in \mathcal{U}$

$$(D \sqcap D)(x) = \max[2D(x) - 1, 0] = 0, \qquad (6.15)$$

and by the condition (d) of the definition of a fuzzy quantum logic

$$D = A \sqcup B - A \vee B = \emptyset. \qquad (6.16)$$

This means that for any $x \in \mathcal{U}$

$$(A \sqcup B)(x) - (A \vee B)(x) = 0, \qquad (6.17)$$

which finishes the proof of part (a).

The proof of part (b) is analogous. $\square$

Proof of Theorem 3. Using the Lemma 3.1 of [9] already mentioned one can easily check that any quantum logic of fuzzy sets is a D-poset in the sense of Kôpka [14] in which a partial order relation is the standard fuzzy set inclusion (4.3) and a general difference that appears in the definition of a D-poset is the pointwise difference of fuzzy sets (6.13).

We shall prove first that if A, B, $A \sqcap B \in \mathcal{L}(\mathcal{U})$, then $A \sqcup B \in \mathcal{L}(\mathcal{U})$.

Let us assume that A, B, $A \sqcap B \in \mathcal{L}(\mathcal{U})$. It can easily be checked that $A - (A \sqcap B) \subseteq B' = \mathcal{U} - B$ and $B - (A \sqcap B) \subseteq A' = \mathcal{U} - A$, and therefore, $A \sqcap B$ fulfills the condition (3) of Theorem 1 of [15]. Consequently, we infer from the condition (4) of this Theorem that there exists a triplet $\{A_1, B_1, C\}$ of pairwise weakly disjoint (orthogonal) elements of $\mathcal{L}(\mathcal{U})$ such that $A = A_1 + C$ and $B = B_1 + C$ (the sums defined pointwisely on $\mathcal{U}$). Moreover, it can be inferred from Kôpka's proof that in our case $C = A \sqcap B$. Therefore, $A_1 + B_1 + C = A - C + B - C + C = A + B - A \sqcap B = A \sqcup B$, and from the condition (c) of the definition of a fuzzy quantum logic it follows that $A \sqcup B \in \mathcal{L}(\mathcal{U})$.

The proof of the implication A, B, $A \sqcup B \in \mathcal{L}(\mathcal{U}) \Rightarrow A \sqcap B \in \mathcal{L}(\mathcal{U})$ also utilizes Theorem 1 of [15] and proceeds analogously.

Since weak disjointness of elements of a quantum logic of fuzzy sets is equivalent to their orthogonality (in the traditional sense), the joins $A_1 \vee C$ and $B_1 \vee C$ exist in $\mathcal{L}(\mathcal{U})$ by condition (d) of the (traditional) definition of a quantum logic. Since for pairwise weakly disjoint sets a membership function of their Łukasiewicz union coincides with an algebraic sum of their membership functions [10], it follows from Lemma 1 and from the condition (c) of the definition of a fuzzy quantum logic that

$$A_1 \vee C = A_1 \sqcup C = A_1 + C = A - C + C = A \tag{6.18}$$

$$B_1 \vee C = B_1 \sqcup C = B_1 + C = B - C + C = B, \tag{6.19}$$

which proves the compatibility of A and B.

Finally, let us notice that any quantum logic contains meets and joins of all pairs of compatible elements. Therefore, it follows from Lemma 1 that

$$A \sqcap B = A \wedge B, \quad A \sqcup B = A \vee B, \tag{6.20}$$

which finishes the proof of Theorem 3.   □

Theorem 3 cannot be "reversed" in the sense that for arbitrary $A, B \in \mathcal{L}(\mathcal{U})$ neither the existence of $A \wedge B$, $A \vee B$ in $\mathcal{L}(\mathcal{U})$, nor the compatibility of A and B implies the existence of $A \sqcap B$, $A \sqcup B$ in $\mathcal{L}(\mathcal{U})$. To justify the first part of this statement let us note that, since by Theorem 3 the existence of a Łukasiewicz union or intersection of any two elements of $\mathcal{L}(\mathcal{U})$ implies their compatibility and since Boolean algebras are quantum logics in which all elements are compatible (see, e.g., [5]), Theorem 3 yields the following Corollary:

Corollary 1 *If $A \sqcap B$ or $A \sqcup B$ exists in a fuzzy quantum logic for all $A, B \in \mathcal{L}(\mathcal{U})$, $A \neq B$, then $\mathcal{L}(\mathcal{U})$ is a Boolean algebra.*

Therefore, if the existence of a meet or join of two elements $A, B \in \mathcal{L}(\mathcal{U})$ had been sufficient for the existence in $\mathcal{L}(\mathcal{U})$ of their Łukasiewicz union and intersection, every orthomodular lattice would have been a Boolean algebra, which is obviously not true. The fact that compatibility of two elements of a fuzzy quantum logic is not sufficient for the existence of their Łukasiewicz union and intersection is shown in the following example given already in [12]. This example actually shows that Corollary 1 also cannot be "reversed", i.e., that even if $\mathcal{L}(\mathcal{U})$ is a Boolean algebra, $A \sqcap B$ and $A \sqcup B$ do not necessarily belong to $\mathcal{L}(\mathcal{U})$ for all possible pairs $A, B \in \mathcal{L}(\mathcal{U})$, $A \neq B$.

Example 1 Let X be a triangle with vertices a, b, c, and let $\mathcal{L}(X)$ consist of all fuzzy subsets of X whose membership functions are affine functions and assume on all vertices of X values 0 or 1 only. It was proved in [16] that such $\mathcal{L}(X)$ is a Boolean algebra, therefore, all its elements are compatible. Let us denote A, B elements of $\mathcal{L}(X)$ such that $A(a) = 1$, $A(b) = A(c) = 0$, $B(b) = 0$, $B(a) = B(c) = 1$, and let us construct their Łukasiewicz union and intersection. Since $(A \sqcup B)(a) = \min(1+1, 1) = 1, (A \sqcup B)(b) = \min(0+0, 1) = 0, (A \sqcup B)(c) = \min(0+1, 1) = 1$, and $(A \sqcap B)(a) = \max(1 + 1 - 1, 0) = 1, (A \sqcap B)(b) = \max(0 + 0 - 1, 0) = 0$, $(A \sqcap B)(c) = \max(0+1-1, 0) = 0$, $A \sqcup B$ and $A \sqcap B$ would have belonged to $\mathcal{L}(X)$ iff their membership functions had been affine. But this is impossible: Let d be the midpoint of the edge ab. Since values taken by affine functions on the point d should be arithmetic means of their values taken on the vertices a and b, it may be expected that $(A \sqcup B)(d) = (A \sqcap B)(d) = \frac{1}{2}$. However, from $A(d) = B(d) = \frac{1}{2}$, it is inferred that $(A \sqcup B)(d) = \min(\frac{1}{2} + \frac{1}{2}, 1) = 1$, and $(A \sqcap B)(d) = \max(\frac{1}{2} + \frac{1}{2} - 1, 0) = 0$. Therefore, membership functions of the sets $A \sqcup B$ and $A \sqcap B$ are not affine functions and the sets $A \sqcup B$ and $A \sqcap B$ do not belong to $\mathcal{L}(X)$.

The other difference between the Łukasiewicz and order-theoretic operations on the quantum logics of fuzzy sets concerns idempotency: Order-theoretic operations of meet and join are idempotent, i.e. $A \wedge A = A \vee A = A$ for any $A \in \mathcal{L}(\mathcal{U})$. In contrast, neither $A \sqcup A$, nor $A \sqcap A$ belongs to $\mathcal{L}(\mathcal{U})$ for any "genuine fuzzy", i.e., non-crisp element of $\mathcal{L}(\mathcal{U})$:

Theorem 4 *Let $\mathcal{L}(\mathcal{U})$ be a quantum logic of fuzzy subsets of a universe $\mathcal{U}$ and let $A \in \mathcal{L}(\mathcal{U})$. Then $A \sqcup A$ and $A \sqcap A$ belong to $\mathcal{L}(\mathcal{U})$ iff A is a crisp subset of $\mathcal{U}$, i.e., iff $A(x) \in \{0, 1\}$ for all $x \in \mathcal{U}$. In this and only in this case Łukasiewicz operations coincide with ordinary set-theoretic union and intersection, and are idempotent.*

Proof Let us note first that Łukasiewicz's operations are never idempotent when applied to a genuine fuzzy subset A of $\mathcal{U}$, i.e., if there exists $x \in \mathcal{U}$ such that $A(x) \neq 0, 1$. Indeed, there are two possibilities:

1. If $0 < A(x) \leq \frac{1}{2}$, then $(A \sqcap A)(x) = \max[2A(x) - 1, 0] = 0 \neq A(x)$, and
   $(A \sqcup A)(x) = \min[2A(x), 1] = 2A(x) \neq A(x)$.
2. If $\frac{1}{2} < A(x) < 1$, then $(A \sqcap A)(x) = 2A(x) - 1$ which, if equal to $A(x)$ would
   imply $A(x) = 1$, and $(A \sqcup A)(x) = 1 \neq A(x)$.

The fact that various unions and intersections of fuzzy sets when applied to crisp sets coincide with ordinary set-theoretic union and intersection (which are obviously idempotent) belongs to mathematical folklore. In particular, the crispness of $A \in \mathcal{L}(\mathcal{U})$ obviously yields

$$A \sqcap A = A \wedge A = A = A \vee A = A \sqcup A \in \mathcal{L}(\mathcal{U}). \tag{6.21}$$

Therefore, the nontrivial part of the proof consists in showing that if $A \sqcap A$ and $A \sqcup A$ belong to $\mathcal{L}(\mathcal{U})$, then A is crisp. This, however, follows from Theorem 3 since in this case the equalities (6.21) hold and, by the first part of the proof, A is necessarily crisp.

$\square$

Before closing this section let us consider an often studied case of a so called *concrete logic* (see, e.g., [5]) which is a family Δ of crisp subsets of a fixed set Ω such that

(a) $\emptyset \in \Delta$
(b) if $A \in \Delta$, then $\Omega - A \in \Delta$
(c) if $\{A_i : i \in \mathbb{N}\}$ is a countable family of mutually disjoint sets, then $\cup_i A_i \in \Delta$.

Since the traditional set-theoretic operations can also be thought of as the Łukasiewicz operations applied to this "degenerate" (from the point of view of the fuzzy set theory) case, we see that all three conditions that appear in the definition of a concrete logic are actually identical with the first three conditions of the definition of a quantum logic of fuzzy sets. Since the remaining condition (d) of the definition of a fuzzy quantum logic (if $A \sqcap A = \emptyset$ then $A = \emptyset$) is always satisfied by any crisp set A, we see that our definition of a quantum logic of fuzzy sets is a straightforward generalization of the definition of a concrete logic or, the other way round, that every quantum logic of fuzzy sets automatically becomes a concrete logic if all its elements are crisp.

Let us, however, note that concrete logics, although providing nice mathematical examples of a general notion of a quantum logic, are not very interesting from the physical point of view, since they admit ordering families of two-valued (i.e., dispersion-free) states while, by Gleason's theorem, Hilbertian quantum logics do not possess any two-valued state if the dimension of a Hilbert space is strictly greater than two (see, e.g., [5]). Therefore, abandoning crisp models and working with genuine fuzzy sets is an indispensable step if one wants to produce set-theoretic models of quantum logics which are of any value to quantum physics.

References

1. Birkhoff, G. and J. von Neumann, "The logic of quantum mechanics", *Annals of Mathematics*, **37** (1936) 823–843.
2. Husimi, K. "Studies on the foundations of quantum mechanics I", *Proceedings of the Physico-Mathematical Society of Japan*, **19** (1937) 766–789.
3. Burmeister, P. and M. Mączyński, "Orthomodular (partial) algebras and their representations", *Demonstratio Mathematica*, **27** (1994) 701–722.
4. Beltrametti, E. G. and G. Cassinelli, *The Logic of Quantum Mechanics* (Addison-Wesley, Reading, 1981).
5. Pták, P. and S. Pulmannová, *Orthomodular Structures as Quantum Logics* (Kluwer Academic Publishers, Dordrecht, 1991).
6. Mączyński, M. J. "The orthogonality postulate in axiomatic quantum mechanics", *International Journal of Theoretical Physics*, **8** (1973) 353–360.
7. Mączyński, M. J. "Functional properties of quantum logics", *International Journal of Theoretical Physics*, **11** (1974) 149–156.
8. Pykacz, J. "Quantum logics as families of fuzzy subsets of the set of physical states", *Preprints of the Second International Fuzzy Systems Association Congress, Tokyo, July 20–25, 1987*, vol **2**, 437–440.
9. Pykacz, J. "Fuzzy set ideas in quantum logics", *International Journal of Theoretical Physics*, **31** (1992) 1767–1783.
10. Pykacz, J. "Fuzzy quantum logics and infinite-valued Łukasiewicz logic", *International Journal of Theoretical Physics*, **33** (1994) 1403–1416.
11. Giles, R. "Łukasiewicz logic and fuzzy set theory", *International Journal of Man-Machine Studies*, **67** (1976) 313–327.
12. Pykacz, J. "Attempts at the logical explanation of the wave-particle duality", in: M. L. Dalla Chiara et al. (eds) *Language, Quantum, Music* (Kluwer, Dordrecht, 1999) 269–282.
13. Pykacz, J. "Łukasiewicz operations in fuzzy set and many-valued representations of quantum logics", *Foundations of Physics*, **30** (2000) 1503–1524.
14. Kôpka, F. "D-posets of fuzzy sets", *Tatra Mountains Mathematical Publications*, **1** (1992) 83–87.
15. Kôpka, F. "Compatibility of D-posets of fuzzy sets", *Tatra Mountains Mathematical Publications*, **6** (1995) 95–102.
16. Pykacz, J. "Affine Mączyński logics on compact convex sets of states", *International Journal of Theoretical Physics*, **22** (1983) 97–106.

Chapter 7
B-vN Quantum Logic as ∞-Valued Łukasiewicz Logic

7.1 The Necessity of Using Many-Valued Logic for Description of Future Non-certain Events

Jan Łukasiewicz argued in his address which he delivered as a Rector of Warsaw University at the Inaguration of the academic year 1922/1923 [1] that *"All sentences about future facts that are not yet decided belong to this* [many-valued] *category. Such sentences are neither true at the present moment ... nor are they false..."*. This attitude is clearly applicable to sentences concerning the results of future experiments which are not decided at present. Let us note the full analogy between Aristotle's statement often quoted by Łukasiewicz *"There will be a sea battle tomorrow"* and quantum mechanical predictions of the form *"A photon will pass through a filter"*. In both cases the position of classical 2-valued logic is such that since the occurrence or non-occurrence of these events is not certain, these statements cannot be endowed with any truth value, i.e., they do not belong to the domain of 2-valued logic. Therefore, changing 2-valued logic to many-valued logic while discussing future non-certain events seems to be inevitable—otherwise we could not reasonably talk about them at all.

Let us note that in fact we use, usually unconsciously, many-valued logic in such situations in everyday life. When one says *"There will be a sea battle tomorrow"*, one usually intuitively ascribes some degree of probability (possibility?, likelihood?) to this future event. This is exactly the way in which Łukasiewicz interpreted truth values of statements about future events that are different form 0 and 1 in his numerous papers [2].[1]

While considering future events in the quantum realm we are in much better situation because these truth values are known with absolute precision: they are

---

[1] The idea that probabilities should be interpreted as truth values of many-valued logic was for the first time expressed by Łukasiewicz already in 1913 in [3].

© The Author(s) 2015
J. Pykacz, *Quantum Physics, Fuzzy Sets and Logic*,
SpringerBriefs in Physics, DOI 10.1007/978-3-319-19384-7_7

simply probabilities that these events will happen, obtained from the theory by exact calculations and the Born rule. Let us note that when an experiment is finished and its results are known with certainty we are back to classical 2-valued logic—exactly the same happens in macroscopic world (sea battle either happened or it did not happen).

In macroscopic world usually the process of converging of many-valued truth values to the extreme values 0 or 1 can be seen "dynamically" and can be traced in a more detailed way: when two hostile navies approach each other, truth value of a sentence *"There will be a sea battle"* tends to unity and finally attains it at the moment the sea battle begins. On the contrary, if navies are more and more distant, truth value of the considered sentence tends to 0. In the quantum realm such detailed description of the process of passing from many-valued to 2-valued logic is not provided, except for Objective Collapse Theories.

7.2 Is B-vN Quantum Logic Two-Valued?

According to the traditional approach, Birkhoff-von Neumann quantum logic defined in the previous chapter is considered to be a 2-valued logic which is, however, non-classical because distributive law is replaced in it by a weaker orthomodular law. The alleged 2-valuedness follows from the fact that an element of a quantum logic is interpreted as an experimentally verifiable "yes-no" proposition about a physical system that occurs to be either true or false when a suitable experiment, designed to check it, is completed. It should be stressed that this alleged 2-valuedness concerns truth-evaluation made post factum, while pre factum truth-evaluation of such statements, according to argumentation originated by Łukasiewicz and presented in the previous section, inevitably forces us to enter the domain of many-valued logic.

Let us also note that elements of a quantum logic can be equivalently thought of as mathematical representations of *properties* of a physical system (e.g. a photon's property of *passing through a polarizer*). In the case of classical physical systems properties of these systems or, equivalently, two-valued propositions that express these properties, define uniquely traditional (crisp) subsets of sets of pure states (phase spaces) of these systems consisting of exactly those pure states for which the considered properties hold. On the other hand, each crisp subset A of a phase space defines a two-valued proposition about a physical system (e.g., of the form *"a state of a system belongs to A"*) which is obviously equivalent to the original one that defines a subset A, so one can go back and forth in this construction without meeting any difficulties.

In the case of a quantum system the construction described above of *crisp* subsets of a phase space that remain in one-to-one correspondence with properties of a physical system cannot, in general, be performed. In most cases even if a quantum object (e.g., a photon) is in a pure state (e.g., in a definite state of linear polarization) one is not able to state whether it has or has not a specific property (e.g. a property of *passing through a linear polarizer oriented under the angle* $\alpha \neq 0, \frac{\pi}{2}$ *to the direction of photon's polarization*) before the suitable experiment is done. It is, in fact, a credo of the Copenhagen Interpretation of quantum mechanics that *"unperformed experiments have no results"* [4]. Consequently, the Copenhagen School forbids even to consider whether a quantum system *has* or *has not* any of its properties before a suitable experiment is performed. This is very much like forbidding to consider whether a navy N is victorious (or not) before the experiment (sea battle) is completed. However, we can often evaluate on the basis of other information (the number of ships, quality of weapon, previous battles, etc.) probability that the navy N will tomorrow win a sea battle. In this sense, according to the very spirit of probabilistic interpretation of truth-values in infinite-valued logic we can say, even before the battle is finished, that the navy N has a property of *being victorious* to the degree, say, 0.8 and at the same time that it has a property of *not being victorious* to the degree $1 - 0.8 = 0.2$ or, equivalently, that the proposition *"the navy N will win a sea battle tomorrow"* is *today* true to the degree 0.8 and false to the degree 0.2.

It is sometimes argued that if we knew, like Laplace's Demon, all initial conditions with absolute accuracy, then the course of all future events could be predicted with certainty, so the apparent randomness, indeterminism, etc., that we see around would disappear and nothing but the classical 2-valued logic would be necessary to describe all events, also the future ones. This argument is false since it does not concern the Real World that we live in but an Imaginary World that, to the best of our contemporary knowledge, does not exist: The Real World we live in is a quantum world so the very possibility of knowing simultaneously, with absolute accuracy, positions and momenta of all particles that form any object (e.g., warships and their crews) is excluded by the fundamental laws of quantum mechanics.

The situation of a linearly polarized photon before it reaches a polarizer is of exactly the same type as the previously considered situation of a navy before a sea battle. According to the "many-valued interpretation" we can safely say that a linearly polarized photon actually possesses a property of *being able to pass through a linear polarizer oriented under the angle* α *to the direction of its polarization* to the degree defined by the Malus Law, i.e. $\cos^2 \alpha$ or, equivalently, that $\cos^2 \alpha$ is the truth value of the proposition *"a linearly polarized photon will pass through a linear polarizer oriented under the angle* α *to the direction of its polarization"*. To paraphrase the title of Peres' paper [4], we can say that *"unperformed experiments have all their possible results, each of them to the degree allowed by suitable quantum-mechanical calculations"*.

Of course when the experiment is completed we are back to the classical two-valued logic since a photon either actually passed or did not pass through the polarizer (similarly, a sea battle either happened or did not happen). Loosely speaking, we can say that during an experiment a kind of a "logical collapse" of the infinite-valued logic onto the two-valued logic occurs, this collapse being a logical counterpart of von Neumann's collapse of a wave function.

7.3 The Many-Valued Model of B-vN Quantum Logic

So far we have established that statements about the results of yet to be performed experiments on quantum systems or about yet to be tested properties of these objects should be treated as belonging to the domain of infinite-valued logic. The natural question arises as to what kind of logical operations and what kind of structure should families of such statements be endowed with? Theorem 2 in Chap. 6, which enables any Birkhoff–von Neumann quantum logic L possessing an ordering set of states S to be expressed as a family $\mathcal{L}(S)$ of fuzzy subsets of S endowed with partially defined Łukasiewicz operations, together with basic relations between fuzzy sets and infinite-valued logic described in Chap. 4, strongly suggest that one small step more should be taken and L should be represented further as a family of infinite-valued propositions. However, before we do this let us note that the identification of a (many-valued) truth-value of a proposition "x *belongs to A*" with a value of a membership function $\mu_A(x)$ was carried out "locally" in the point x , so the whole fuzzy set A is in one-to-one correspondence not with a single proposition but with a *propositional function* $a(\cdot) = $ "$\cdot$ *belongs to A*". This propositional function becomes a proposition, i.e. it may be endowed with a truth value only when we insert into it a (name of a) specific element of a considered universe of discourse, in our case a (name of a) specific state of the considered physical system.

There are, in general, constant propositional functions (therefore, they are, in fact, propositions) that assume the same truth value for every argument in their domain. We shall be particularly interested in two of them: the *always-false propositional function* f that assumes truth value 0 in all points of its domain and the *always-true propositional function* t that assumes truth value 1 in all points of its domain. In the case of propositional functions about physical systems they can be linguistically expressed, for example, as $f = $ "*the studied physical system does not exist*" and $t = $ "*the studied physical system exists*", but other linguistic expressions for these propositional functions are also possible. However, it follows from the results concerning fuzzy set representation of the B-vN quantum logic that many-valued representation of the B-vN quantum logic does not allow any constant propositional function except f and t (we quote here from [5] the relevant theorem in its original fuzzy set formulation):

Theorem 5 *A quantum logic* $\mathcal{L}(\mathcal{U})$ *of fuzzy subsets of a universe* $\mathcal{U}$ *does not contain any set whose membership function is constant except crisp sets* $\emptyset$ *and* $\mathcal{U}$.

Proof If $A \in \mathcal{L}(\mathcal{U})$ has a constant membership function, then either $A \subseteq A'$, or $A' \subseteq A$. In the first case $A \wedge A' = A$ which, by the condition (c) of the traditional definition of a quantum logic, means that $A = \emptyset$. In the second case $A \vee A' = A$ which, by the same condition, means that $A = \mathcal{U}$. □

In what follows we shall always be dealing with families of propositional functions defined on common domains. In such cases one can define logical operations on propositional functions in a pointwise way, for example

$$c(\cdot) \equiv a(\cdot) \sqcap b(\cdot) \text{ iff } c(x) \equiv a(x) \sqcap b(x), \tag{7.1}$$

for all arguments x in their common domain, where the symbol $\equiv$ denotes "has the same truth value as", i.e. for any argument x in the common domain of these propositional functions

$$\tau(c(x)) = \max[\tau(a(x)) + \tau(b(x)) - 1, 0]. \tag{7.2}$$

One can easily note that if the always-false and always-true propositional functions are defined on the common domain, then each of them is the Łukasiewicz negation (3.1) of the other.

Two propositional functions $a(\cdot), b(\cdot)$ (defined on the common domain) will be termed *exclusive* or we shall say that they *exclude each other* iff their Łukasiewicz conjunction (3.13) is always false, i.e., iff

$$a(\cdot) \sqcap b(\cdot) \equiv f \tag{7.3}$$

which, by (7.2), means that

$$\tau(a(x)) + \tau(b(x)) \le 1 \tag{7.4}$$

for any argument x in their common domain.

Now we can reformulate Theorem 2 in Chap. 6 using the language of many-valued logic and the above-introduced notion of exclusive propositional functions.

Theorem 6 *Any quantum logic L with an ordering set of probability measures S can be isomorphically represented as a family* $\mathbb{L}(S)$ *of propositional functions defined on S and satisfying the following conditions:*

(a) $\mathbb{L}(S)$ *contains the always-false propositional function* f.
(b) $\mathbb{L}(S)$ *is closed with respect to the Łukasiewicz negation (3.6), i.e., if* $a(\cdot) \in \mathbb{L}(S)$, *then* $\neg a(\cdot) \in \mathbb{L}(S)$.
(c) $\mathbb{L}(S)$ *is closed with respect to the Łukasiewicz disjunction (3.12) of pairwise exclusive propositional functions, i.e. if* $a(\cdot)_i \sqcap a(\cdot)_j \equiv f$ *for* $i \ne j$, *then* $\sqcup_i a(\cdot)_i \in \mathbb{L}(S)$.
(d) *The always-false propositional function* f *is the only propositional function in* $\mathbb{L}(S)$ *that excludes itself, i.e., for any* $a(\cdot) \in \mathbb{L}(S)$, *if* $a(\cdot) \sqcap a(\cdot) \equiv f$, *then* $a(\cdot) \equiv f$.

Conversely, any family of many-valued propositional functions defined on a common domain D and satisfying conditions (a)–(d) is a quantum logic in the Birkhoff–von Neumann sense whenever we identify propositional functions that assume the same truth value for every argument in their common domain D. This family is partially ordered by the partial order relation generated by the Łukasiewicz implication (3.5):

$$a(\cdot) \le b(\cdot) \text{ iff } [a(\cdot) \to b(\cdot)] \equiv t, \tag{7.5}$$

with the Łukasiewicz negation (3.6) as ortho complementation, the orthogonality
of the elements coinciding with their exclusiveness (7.3), and an ordering set of
probability measures being generated by arguments in their common domain D
according to the formula

$$s_x(a(\cdot)) = \tau(a(x)) \text{ for all } x \in D. \tag{7.6}$$

Proof Let us note that the right-hand side of the formula (7.5) means that for every
$x \in D$

$$\tau[a(x) \to b(x)] = \min[1 - \tau(a(x)) + \tau(b(x)), 1] = \tau(t) = 1, \tag{7.7}$$

which yields

$$\tau(a(x)) \le \tau(b(x)) \tag{7.8}$$

for any $x \in D$, i.e., it actually establishes pointwise partial order between propositional functions, provided that we identify propositional functions that assume the
same truth value on all elements in their common domain, which is a standard procedure in the construction of the Lindenbaum algebras of logics.

In view of the links between fuzzy sets and infinite-valued logic, which have
been extensively commented on, all the remaining conditions of Theorem 6 are just
"many-valued translations" of respective conditions of Theorem 2 in Chap. 6, which
finishes the proof. □

Let us see now what can be said about the "many-valued model" $\mathbb{L}(S)$ of the B-vN
quantum logic, in particular about the Łukasiewicz logical operations that appear in
it, in view of the results already obtained for the "fuzzy set model" $\mathcal{L}(S)$ of this logic.

We infer from Corollary 1 that $\mathbb{L}(S)$ is, in general, only a partial logic, i.e. the
Łukasiewicz conjunction ⊓ and disjunction ⊔ are *not* defined for all pairs of propositional functions, since if they are, $\mathbb{L}(S)$ necessarily has an algebraic structure of
a Boolean algebra. Let us note that in view of the widely discussed (see, for example, Chap. 8 of Jammer's book [6]) interpretational difficulties in treating the order-
theoretic operations of meet ∧ and join ∨ as proper models of conjunction and disjunction in the case of noncompatible elements of an orthomodular lattice of closed
subspaces of a Hilbert space, the fact that ⊓ and ⊔ are only partially defined on $\mathbb{L}(S)$
is more of a virtue than a drawback. On the other hand Theorem 3 in Chap. 6 implies
that Łukasiewicz's connectives are properly modelled by the order-theoretic operations whenever the former are defined. Therefore, if we adopt the hypothesis that ⊓
and ⊔, *rather than* ∧ *and* ∨, *are proper models of the quantum-logical conjunction
and disjunction*, Theorem 3 in Chap. 6 explains why Birkhoff and von Neumann (and
their followers during 60 years of development of quantum logic theory) could treat
meet and join as algebraic representations of the quantum-logical conjunction and
disjunction in spite of the difficulties raised by this choice. It should be mentioned,
however, that in view of Example 1 (the non-existence of Łukasiewicz's operations
on some pairs of elements of some Boolean algebras) and also in view of the general

non-idempotency of these operations, the problem of the proper "logical" interpretation of $\sqcap$ and $\sqcup$ is far from being settled.

The fact that $a(\cdot) \sqcup b(\cdot)$ is defined when $a(\cdot) \sqcap b(\cdot) \equiv f$ may suggest that the Łukasiewicz disjunction $\sqcup$ is, in the case of quantum logics, a many-valued counterpart of a two-valued *exclusive disjunction* (*exclusive-OR, XOR*) rather than a counterpart of the ordinary disjunction. However, this hypothesis is based on the tacit assumption that $\sqcap$ is a many-valued counterpart of the ordinary conjunction, while a two-valued connective associated to the exclusive-OR by de Morgan's laws is the equivalence, not conjunction. Moreover, the case $a(\cdot) \sqcap b(\cdot) \equiv f$ is not the only case when $a(\cdot) \sqcap b(\cdot)$ and $a(\cdot) \sqcup b(\cdot)$ are defined in $\mathbb{L}(S)$. Indeed, if $\tau(a(x)) + \tau(b(x)) \geq 1$ for all $x \in S$, which means that $a(\cdot) \sqcup b(\cdot) \equiv t$, then $\tau(\neg a(x)) + \tau(\neg b(x)) = 1 - \tau(a(x)) + 1 - \tau(b(x)) \leq 1$, so $(\neg a(\cdot)) \sqcap (\neg b(\cdot)) \equiv f$ and by Theorem 6 (c) $(\neg a(\cdot)) \sqcup (\neg b(\cdot)) \in \mathbb{L}(S)$, which, by condition (b) of this theorem, de Morgan's identity, and the many-valued version of Theorem 3 in Chap. 6 yields $\neg[(\neg a(\cdot)) \sqcup (\neg b(\cdot))] \equiv a(\cdot) \sqcap b(\cdot) \in \mathbb{L}(S)$, although in this case the Łukasiewicz conjunction $a(\cdot) \sqcap b(\cdot)$ is not always false.

Let us note that since by conditions (a) and (b) of Theorem 6 the always-true propositional function t belongs to $\mathbb{L}(S)$, the fact that $a(\cdot) \sqcup b(\cdot) \equiv t$ implies $a(\cdot) \sqcap b(\cdot) \in \mathbb{L}(S)$ is also a straightforward consequence of Theorem 3 in Chap. 6.

We shall show now that the two possibilities listed above are the only possibilities allowed for genuine many-valued propositional functions that belong to $\mathbb{L}(S)$ when a further regularity condition stating that these functions should behave well with respect to convex mixtures of states (i.e., be affine functions) is assumed. This means that the Łukasiewicz conjunction $\sqcap$ and disjunction $\sqcup$ are defined for a pair of propositional functions $a(\cdot), b(\cdot) \in \mathbb{L}(S)$ only when their Łukasiewicz conjunction $a(\cdot) \sqcap b(\cdot)$ is always false or when their Łukasiewicz disjunction $a(\cdot) \sqcup b(\cdot)$ is always true. This fact will be shown in the following theorem which, for technical simplicity, is again presented in the equivalent fuzzy set formulation.

Theorem 7 *Let $\mathcal{U}$ be a convex set and let $\mathcal{L}(\mathcal{U})$ be a quantum logic of fuzzy subsets of $\mathcal{U}$. If all elements of $\mathcal{L}(\mathcal{U})$ have affine membership functions, then for any $A, B \in \mathcal{L}(\mathcal{U})$, $A \sqcap B$ and $A \sqcup B$ belong to $\mathcal{L}(\mathcal{U})$ if and only if $A \sqcap B = \emptyset$ or $A \sqcup B = \mathcal{U}$.*

Proof The "if" part of the proof follows immediately from the condition (c) of the definition of a quantum logic of fuzzy sets and considerations contained in the last but one paragraph preceding this theorem.

In order to prove the "only if" part let us recall that $A \sqcap B = \emptyset$ means that $A(x) + B(x) \leq 1$ for all $x \in \mathcal{U}$ while $A \sqcup B = \mathcal{U}$ means that $A(x) + B(x) \geq 1$ for all $x \in \mathcal{U}$. Let us assume that neither of these conditions hold, i.e. that there exist $x_1, x_2 \in \mathcal{U}$ such that $A(x_1) + B(x_1) = (A + B)(x_1) > 1$ and $A(x_2) + B(x_2) = (A + B)(x_2) < 1$ which, respectively, means that $(A \sqcup B)(x_1) = 1$ and $(A \sqcap B)(x_2) = 0$. Since the sum of affine functions is also an affine function, this implies that an interval $(x_1, x_2) = \{x \mid x = \alpha x_1 + (1 - \alpha)x_2, \alpha \in (0, 1)\}$ contains a point $y = \beta x_1 + (1 - \beta)x_2, \beta \in (0, 1)$ such that $(A + B)(y) = A(y) + B(y) = 1$. This makes $A \sqcap B \in \mathcal{L}(\mathcal{U})$ and $A \sqcup B \in \mathcal{L}(\mathcal{U})$ impossible since their membership functions cannot be affine: indeed, if they were affine, then $(A \sqcap B)(y) = \beta(A \sqcap B)(x_1) + (1 - \beta)(A \sqcap B)(x_2) = \beta[A(x_1) + B(x_1) - 1] + (1 - \beta) \cdot 0 > 0$ and $(A \sqcup B)(y) = \beta(A \sqcup B)(x_1) + (1 - \beta)(A \sqcup B)(x_2) = \beta \cdot 1 + (1 - \beta)[A(x_2) + B(x_2)] = \beta[1 - (A + B)(x_2)] + (A + B)(x_2) < 1 - (A + B)(x_2) + (A + B)(x_2) = 1.\ \square$

From the physical point of view the assumption of the affinity of all propositional functions that express properties of a physical system is very natural since it guarantees that these functions behave "correctly" with respect to decompositions of mixed states into their pure components. The following example shows that without this assumption Theorem 7 could not be proved.

Example 2 Let $\mathcal{U} = \{x_1, x_2, x_3, x_4\}$ and let $\mathcal{L}(\mathcal{U})$ consist of fuzzy subsets of $\mathcal{U}$ whose membership functions take in these points the respective values:

$$\emptyset = \{0, 0, 0, 0\} \qquad\qquad \mathcal{U} = \{1, 1, 1, 1\}$$
$$A = \{0.7, 0, 0, 0\} \qquad\qquad A' = \{0.3, 1, 1, 1\}$$
$$B = \{0, 0.7, 0, 0\} \qquad\qquad B' = \{1, 0.3, 1, 1\}$$
$$C = \{0.1, 0.2, 0, 1\} \qquad\quad C' = \{0.9, 0.8, 1, 0\}$$
$$D = \{0.2, 0.1, 1, 0\} \qquad\quad D' = \{0.8, 0.9, 0, 1\}$$
$$E = \{0.1, 0.9, 0, 1\} \qquad\quad E' = \{0.9, 0.1, 1, 0\}$$
$$F = \{0.2, 0.8, 1, 0\} \qquad\quad F' = \{0.8, 0.2, 0, 1\}$$
$$G = \{0.3, 0.3, 1, 1\} \qquad\quad G' = \{0.7, 0.7, 0, 0\}$$

It can be checked that $\mathcal{L}(\mathcal{U})$ is a quantum logic of fuzzy subsets of $\mathcal{U}$. Actually, it is isomorphic to 2^4, i.e. to the Boolean algebra of all crisp subsets of $\mathcal{U}$. Since $E \sqcap F = \{0, 0.7, 0, 0\} = B \neq \emptyset$ and $E \sqcup F = \{0.3, 1, 1, 1\} = A' \neq \mathcal{U}$, we see that in this case Theorem 7 does not hold.

Let us note that, since in Example 2 not all Łukasiewicz unions and intersections of elements of $\mathcal{L}(\mathcal{U})$ belong to $\mathcal{L}(\mathcal{U})$ (e.g., $E \sqcap B = \{0, 0.6, 0, 0\} \notin \mathcal{L}(\mathcal{U})$, $E \sqcup B = \{0.1, 1, 0, 1\} \notin \mathcal{L}(\mathcal{U})$), the Boolean algebra of fuzzy sets studied in this example can replace that studied in Example 1 in Chap. 6, in order to show that Corollary 1 cannot be reversed and that the compatibility of two elements of a quantum logic of fuzzy sets is not a sufficient condition for the existence of their Łukasiewicz union and intersection in this logic. Finally, it should be mentioned that, since in the case of crisp sets Łukasiewicz operations coincide with the ordinary set-theoretic operations, the same conclusions as drawn from Example 2 could be drawn from considering any Boolean algebra of all crisp subsets of an n-element set with $n \geq 3$. However, $\mathcal{L}(\mathcal{U})$ studied in Example 2 consists of genuine fuzzy sets which ensures that the conclusions obtained are not artifacts caused by using non-fuzzy sets.

Let us finish this section with three general remarks.

(1) Since infinite-valued logic (respectively, fuzzy set theory) is "infinitely many times" more rich in binary operations than two-valued logic (resp. classical set theory), it may be so that any particular pair of them would exhibit some features that are counterintuitive from the point of view of classical two-valued logic (resp. classical set theory). Therefore, it is possible that any particular answer given to the problem of which two-valued connectives are the most proper two-valued counterparts

of the Łukasiewicz operations on $\mathbb{L}(S)$ will give rise to objections based on some counterintuitive consequences of that choice.

(2) In spite of the above-mentioned objections, we stress that the very possibility that quantum-logical conjunction and disjunction are modelled by Łukasiewicz operations rather than by meet and join opens new interesting possibilities and may have far-reaching consequences. In particular, the problems connected with EPR-type experiments, Bell–Kochen-Specker theorem, etc., where always *conjunctions* of propositions are considered (e.g. of the form "*a quantum object in a state s has property a and property b*"), should be carefully reconsidered in view of the possibility that these conjunctions are modelled by Łukasiewcz conjunctions that do not always have to be defined, not by meets that are, in the case of lattices, defined for all possible pairs of propositions.

(3) One of the objections raised against Birkhoff-von Neumann quantum logic (esp. by logicians) is that it is not truth-functional, i.e. that a knowledge of the truth values of elementary statements does not enable the truth values of compound statements to be calculated, in particular conjunctions and disjunctions of elementary statements. Indeed, order-theoretic operations of meet and join are not "local" in the sense that in order to find $a \wedge b$ and $a \vee b$ one must know the whole order-theoretic structure of a poset or a lattice, i.e. the whole relation of partial order defined on a family of propositions. On the contrary, only a knowledge of the truth values of propositions $a(x)$ and $b(x)$ is necessary to calculate truth values of $(a \sqcap b)(x)$ and $(a \sqcup b)(x)$. Thus, Birkhoff–von Neumann quantum logic in its infinite-valued representation turns out, at least in this restricted sense, to be truth-functional.

7.4 Application: Analysis of a Two-Slit Experiment

In order to show the usefulness of the obtained infinite-valued representation of B-vN quantum logic, we shall apply it to the description of a two-slit experiment.

Let us note that since expressions for truth-values of both Łukasiewicz conjunction (3.13) and Łukasiewicz disjunction (3.12) contain the sum $\tau(a) + \tau(b)$, we can distinguish the following four cases:

(1a) If $0 \leq \tau(a) + \tau(b) < 1$, then $\tau(a \sqcap b) = 0$ and $0 \leq \tau(a \sqcup b) < 1$.
(1b) If $\tau(a) + \tau(b) = 1$, then $\tau(a \sqcap b) = 0$ and $\tau(a \sqcup b) = 1$.
(2a) If $1 < \tau(a) + \tau(b) < 2$, then $0 < \tau(a \sqcap b) < 1$ and $\tau(a \sqcup b) = 1$.
(2b) If $\tau(a) + \tau(b) = 2$ (i.e. $\tau(a) = \tau(b) = 1$), then $\tau(a \sqcap b) = 1$ and $\tau(a \sqcup b) = 1$.

Therefore, we obtain the following theorem:

Theorem 8 *For any two infinite-valued propositions a, b either*

(1a) their Łukasiewicz conjunction is false, i.e. $\tau(a \sqcap b) = 0$, and their Łukasiewicz disjunction is not true, i.e. $\tau(a \sqcup b) < 1$, or
(1b) their Łukasiewicz conjunction is false, i.e. $\tau(a \sqcap b) = 0$, and their Łukasiewicz disjunction is true, i.e. $\tau(a \sqcup b) = 1$, or

*(2a) their Łukasiewicz conjunction is neither false nor true, i.e. $0 < \tau(a \sqcap b) < 1$,
      and their Łukasiewicz disjunction is true, i.e. $\tau(a \sqcup b) = 1$, or*
(2b) both their Łukasiewicz conjunction and disjunction are true.

Let us note that, besides the trivial possibility $\tau(a) = \tau(b) = 0$ which obviously yields $\tau(a \sqcap b) = \tau(a \sqcup b) = 0$, only possibilities (1b) and (2b) are allowed by classical bivalent logic. Therefore, infinite-valued logic allows the behaviour of quantum systems to be described more precisely than can be done with the use of classical two-valued logic. We shall use this opportunity to cast some light on the famous wave-particle duality exhibited in the two-slit experiment. Let $a(s)$ (resp. $b(s)$) denote a proposition "*a quantum object in a state s passes through slit A (resp. B)*". If we interpret Łukasiewicz conjunction $a(s) \sqcap b(s)$ as representing a compound proposition "*a quantum object in a state s passes through slit A* and *through slit B*" and Łukasiewicz disjunction $a(s) \sqcup b(s)$ as representing a compound proposition "*a quantum object in a state s passes through slit A* or *through slit B*", then truth-values of these compound propositions can be used to decide whether a quantum object in the state s reveals its wave-like or particle like properties:

- $\tau(a(s) \sqcap b(s)) = 0$ and $\tau(a(s) \sqcup b(s)) = 1$ means that it is impossible to state experimentally that a quantum object in a state s passes simultaneously through slit A *and* through slit B, although we can be sure that in any experiment it will pass *either* through slit A *or* through slit B (i.e. this *or* is exclusive). Therefore, an object in a state s reveals purely particle-like behaviour.
- if $\tau(a \sqcap b) = 1$ (in which case also $\tau(a \sqcup b) = 1$), we can be sure that an object passes simultaneously through slit A *and* through slit B, and so behaves like a pure wave.

These two cases, corresponding to possibilities (1b) and (2b) of Theorem 8, can equally well be described with the aid of classical bivalent logic. However, Theorem 8 also allows possibilities (1a) and (2a) which, in the case of the two-slit experiment, can be interpreted as follows:

(1a) We are not sure whether an object passes through slit A *or* B, although we are sure that it does not pass simultaneously through A *and* through B. This possibility seems to correspond to a non-perfect experiment in which an object behaving like a pure particle can pass either through slit A *or* through slit B (but, because $\tau(a \sqcap b) = 0$, not simultaneously through both slits, so this *or* is again exclusive), although we can not be sure whether it passes through either slit.

(2a) We can be sure that an object passes through slit A *or* through slit B (or through both slits since this *or* is non-exclusive) which allows both wave-like and particle-like behaviour. However, in contrast to the case (2b) in which an object behaves like a pure wave we cannot be sure that it passes through both slits simultaneously (since $\tau(a \sqcap b) < 1$), therefore, its behaviour is not 100% wave-like. On the other hand its behaviour is also not 100% particle-like since, in contrast to the case (1b) we are also not sure that it does not pass through both slits simultaneously (since $\tau(a \sqcap b) > 0$). Therefore, an object reveals partial wave-like and partial particle-like behaviour, i.e. it behaves simultaneously but not "to the full extent" like a wave and like a particle. This kind of behaviour is not allowed by the orthodox Copenhagen paradigm based on classical bivalent logic. Therefore, it came as something of a surprise when Wooters and Zurek [7] demonstrated that quantum mechanical formalism enables non-perfect knowledge to be obtained in a two-slit experiment about the path of a quantum object which shows its non-100 % particle-like behaviour simultaneously with a non-perfect interference pattern, which demonstrates that this behaviour is also non-100 % wave-like. The predictions of Wooters and Zurek were confirmed by Mittelstaedt et al. [8] in a carefully performed experiment in which they found that "A photon possesses simultaneously particle properties and wave properties. ...Even for a very high value of the particle property ... there is still a nonvanishing amount of the wave property ... which leads to an observable and significant interference pattern". This experimental result strongly supports the considerations of the previous section, according to which the proper logic of a quantum system is a version of infinite-valued Łukasiewicz logic endowed with Łukasiewicz negation (3.6), implication (3.5) and partially defined Łukasiewicz disjunction (3.12) and conjunction (3.13).

References

1. Łukasiewicz, J. *An Address Delivered as a Rector of the Warsaw University at the Inauguration of the Academic Year 1922/1933;* reprinted as: "On determinism" in [2], pp. 110–128.
2. Łukasiewicz, J. *Selected Works,* ed. by L. Borkowski (North-Holland, Amsterdam, and PWN—Polish Scientific Publishers, Warszawa, 1970).
3. Łukasiewicz, J. *Die logischen Grundlagen der Wahrscheinlichkeitsrechnung* (Acad. der Wiss. Kraków, 1913); reprinted as "Logical foundations of probability theory" in [2], pp. 16–63.
4. Peres, A. "Unperformed experiments have no results", *American Journal of Physics,* **46** (1978) 745–747.
5. Pykacz, J. "Fuzzy set ideas in quantum logics", *International Journal of Theoretical Physics,* **31** (1992) 1767–1783.
6. Jammer, M. *The Philosophy of Quantum Mechanics* (Wiley-Interscience, New York, 1974).
7. Wooters, W. K. and W. H. Zurek, "Complementation in the double-slit experiment: Quantum nonseparability and a quantitative statement of Bohr's principle", *Physical Reviews D,* **19** (1979) 473–484.
8. Mittelstaedt, P., A. Prieur, and R. Schieder, "Unsharp particle-wave duality in a photon split-beam experiment", *Foundations of Physics,* **17** (1987) 891–903.

Chapter 8
Perspectives

Isomorphic representation of Birkhoff–von Neumann quantum logics, and therefore also of orthomodular lattices $L(\mathcal{H})$ of (orthogonal projections onto) closed subspaces of Hilbert spaces by families of fuzzy sets endowed with Łukasiewicz operations opens new opportunities for solving at least two long-standing problems, namely the development of quantum probability calculus in a way completely analogous to the orthodox Kolmogorovian probability theory, and the construction of a phase space representation of quantum mechanics not plagued by the appearance of negative probabilities. However, it should be stressed that what we present here is only a brief prospect for future studies which will certainly require much further investigation.

8.1 Fuzzy Set Models of Quantum Probability

In some experiments on quantum systems the relative frequencies of obtaining various results, interpreted as probabilities, do not fulfil the numerical constraints imposed by classical (Kolmogorovian) probability theory. Such instances, usually connected with the violation of Bell's inequalities, strongly indicate the necessity of modification of the probability calculus used in quantum mechanics.

There are several approaches to the subject that can generally be termed "quantum probability" and even the brief review of all of these would lead us far beyond the scope of this section. Therefore, we shall concentrate on the quantum-logical treatment of this subject.

In the quantum logic approach to the foundations of quantum mechanics the Kolmogorovian triple $(\Omega, \mathcal{F}, P)$ consisting of a space of elementary events Ω, a Boolean σ-algebra $\mathcal{F}$ of selected subsets of Ω (random events), and a probability measure P, is replaced by a couple (L, p) consisting of a σ-orthocomplete orthomodular poset (i.e. quantum logic) L and a probability measure (state) p defined on L. It follows from the very definition (see Sect. 6.1) that probability measures on quantum logics satisfy all numerical constraints imposed on Kolmogorovian probability measures: they are nonnegative, normalized, and σ-additive on families of

© The Author(s) 2015
J. Pykacz, *Quantum Physics, Fuzzy Sets and Logic*,
SpringerBriefs in Physics, DOI 10.1007/978-3-319-19384-7_8

pairwisely disjoint (in the language of "orthodox" quantum logics: pairwisely orthogonal) elements. However, this clearly does not mean that Kolmogorovian probability calculus, which is based on Boolean σ-algebras, is an adequate tool for quantum mechanics.[1] This is particularly evident in the quantum logic approach where several theorems were proved showing that various versions of Bell-type inequalities are satisfied by probability measures defined on a quantum logic if this logic is a Boolean algebra (see, e.g. papers by Santos [2], Pulmannová and Majernik [3], or Beltrametti and Mączyński [4, 5]).

As well as these numerical and "structural" differences between classical and quantum probabilities there is one more important difference: quantum random events are not subsets of the space of elementary events but mathematical objects of another kind. In the Hilbert space model they are represented by closed subspaces of a Hilbert space (or orthogonal projections onto such subspaces), while in an abstract model they are simply elements of an orthomodular poset. This does not allow quantum random events to be treated as subsets of a phase space of a physical system.

In the quantum logic approach the states of any physical system are represented by probability measures on a logic of this system, and they form a convex set whose extreme points represent the pure states of the system. In the particular case of the phase space description of a classical statistical system its logic is identified with a Boolean σ-algebra of Borel subsets of a phase space and pure states are Dirac measures concentrated on one-point subsets of a phase space, so they may be identified with points of a phase space of a system. On the other hand elements of a logic, i.e. Borel subsets of a phase space may be identified with random events since each random event is in an obvious way defined by a property of the physical system: it consists of those pure states for which the given property holds. Therefore, traditional set-theoretic unions and intersections of random events are generated by disjunctions and conjunctions of propositions about the physical system under study in full accordance with the spirit of Kolmogorovian probability theory.

This is no more true for a quantum system. Properties of a quantum system, represented by the elements of a logic, can not be further represented by crisp subsets of the set of pure states. However, we have shown in Sect. 6.3 that there is a possibility of representing the elements of a logic even of a "genuine" quantum system by *fuzzy* subsets of the set of its pure states. It should be noted that the conditions (a)–(d) that define a quantum logic of fuzzy sets show remarkable similarity to the conditions that define Boolean σ-algebras of random events in the Kolmogorovian probability theory. The difference between the condition (c) and the Kolmogorovian requirement that a σ-algebra of random events should be closed with respect to countable unions of arbitrary, not only pairwise disjoint, sets seems to be unimportant since this requirement of Kolmogorov is superfluous: probability measures are assumed to

---

[1] Ballentine's [1] conviction that he has "refuted any and all claims that 'classical' probability theory is not valid in quantum mechanics" seems to be based on a superficial analysis in which he took into account neither Bell-type inequalities, nor the differences in structures on which classical and quantum probability measures are defined.

be σ-additive on pairwise disjoint, not arbitrary sequences of sets and it is possible to construct reasonable "classical" probability theory with this requirement being suitably modified (for a detailed discussion of this problem see the book of Fine [6]). Since the condition (d) in the domain of crisp sets is trivially satisfied, we infer that a notion of a quantum logic of fuzzy sets is in a sense a "minimal" generalization of the notion of a σ-algebra of random events to a family of fuzzy sets endowed with Łukasiewicz connectives, which enables a reasonable probability calculus to be constructed.

After replacing the abstract quantum logics that appear in the foundations of quantum probability calculus by families of fuzzy subsets of sets of pure states of quantum systems, one obtains, at the price of allowing fuzzy sets to come into play, a perfect parallelism between Kolmogorovian probability calculus applied to classical statistical systems and quantum probability calculus applied to quantum systems: in both cases random events are represented by subsets of sets of pure states of physical systems and they are defined by the properties of these systems. Conjunctions and disjunctions of properties of physical systems define intersections and unions of respective subsets. However, as we argued in Chap. 6, in contrast to the situation encountered in classical statistical physics, in quantum physics the results of these operations do not always belong to a quantum logic of fuzzy sets, even if this logic is a lattice. Therefore, the usage of joins and meets in order to construct "compound" quantum random events—a common practice in quantum probability—instead of Łukasiewicz unions and intersections, can be a source of serious difficulties.

Finally, it should be mentioned that it is possible to build a fuzzy probability theory using, instead of Łukasiewicz operations, other operations chosen from the vast family of fuzzy unions and intersections. This has in fact been done in a number of papers (see, e.g. [7–11] to mention a few) in which a fully-fledged fuzzy probability theory was developed. However, in the majority of these papers their authors use the original Zadeh operations which cannot be used to build fuzzy set models of quantum logics since, as it was earlier noticed (in the realm of a many-valued logic) by Gonseth [12], when combined with the standard fuzzy set complementation, they do not satisfy the excluded middle law and the law of contradiction for any genuine fuzzy set.

8.2 Fuzzy Phase Space Representation of Quantum Mechanics

The standard example of a "genuine" quantum logic (i.e. logic that is non-Boolean and can be used to describe genuine quantum systems) is a *Hilbertian quantum logic* $L(\mathcal{H})$ consisting of closed subspaces of a Hilbert space $\mathcal{H}$ used to describe a quantum system or, equivalently, orthogonal projections onto these closed subspaces. Probability measures on $L(\mathcal{H})$ are generated by density operators via the formula

$$p_{\hat{\rho}}(\hat{A}) = Tr(\hat{\rho}\hat{A}), \tag{8.1}$$

where $\hat{\rho}$ is a density operator representing a state of a physical system and $\hat{A}$ an orthogonal projector. Isomorphic representation, provided by Theorem 2 in Chap. 6, of the Hilbertian quantum logic by a family of fuzzy subsets of the set of density operators could be a first step toward constructing a phase space representation of quantum mechanical systems free from the well-known difficulties connected with the appearance of negative probabilities.

The representation of elements of an abstract quantum logic L by a family of fuzzy subsets of the set of its states enables the logics of quantum systems and the logics of classical statistical systems to be compared more easily, which may provide hints for constructing phase space representations of quantum systems. Both similarities and characteristic differences between these two kinds of logics are particularly well-seen when we restrict the underlying universes on which logics are built to sets P consisting of pure states only. In both cases each property $a \in L$ of a physical system Σ defines, by the formula (6.3), a subset $A \subseteq P$ *consisting of pure states in which the system Σ has the property a* (in other words, the set A is defined by the predicate "has the property a"). In the case of classical statistical systems all subsets of P defined in this way are necessarily traditional crisp sets, since pure states in classical mechanics are dispersion-free: $A(s) = s(a) \in \{0, 1\}$, which expresses the fact that a classical system in a pure state either definitely has or definitely has not any of its properties. Therefore, the membership function of the set $A \subseteq P$ is, in this case, a characteristic function and the set A is crisp.

This is no longer the case in quantum mechanics since here even pure states are, in general, dispersive, so the set $A \subseteq P$ defined in the manner described above is, in general, a genuine fuzzy set. Nevertheless, if we assume that properties of a physical system form a (quantum) logic, in both cases the family $\mathcal{L}(P)$ consisting of all fuzzy subsets of P defined in the above-described manner obviously has to satisfy conditions (a)–(d) of Theorem 2 in Chap. 6. As we noticed in the previous section, in the phase space description of a classical statistical system $\mathcal{L}(P)$ can be identified with a Boolean σ-algebra $\mathcal{B}(\Gamma)$ of Borel subsets of a phase space Γ since it is believed that any such subset represents a property of a classical system. In the Hilbert space description of a quantum system $\mathcal{L}(P)$ can be identified with a family $\mathcal{L}(S^1(\mathcal{H}))$ of fuzzy subsets of the unit sphere $S^1(\mathcal{H})$ in a Hilbert space $\mathcal{H}$ associated with a system. In this case the fuzzy sets $A \subseteq S^1(\mathcal{H})$ which form the quantum logic $\mathcal{L}(S^1(\mathcal{H}))$ are defined by the formula:

$$A(\psi) = \langle \psi | \hat{A} \psi \rangle \tag{8.2}$$

where $\psi \in S^1(\mathcal{H})$ is a unit vector and $\hat{A}$ is an orthogonal projection in $\mathcal{H}$. However, it should in general also be possible to represent the properties of a quantum system by a family of fuzzy subsets of a phase space Γ instead of fuzzy subsets of a unit sphere $S^1(\mathcal{H})$ of a Hilbert space $\mathcal{H}$, obtaining in this way a phase space representation of a quantum system. Such representation could be obtained by mapping points $\psi \in S^1(\mathcal{H})$ onto points $(\langle p \rangle_\psi, \langle q \rangle_\psi) \in \Gamma$ with $\langle p \rangle_\psi, \langle q \rangle_\psi$ being the mean values of the momentum and the position operators in a state ψ respectively. A value $A(\langle p \rangle_\psi, \langle q \rangle_\psi)$

of a membership function of a fuzzy subset $A \subset \Gamma$ that represents a property a should in this case be given by the formula (8.2), i.e.

$$A(\langle p \rangle_\psi, \langle q \rangle_\psi) = \langle \psi | \hat{A} \psi \rangle \tag{8.3}$$

where $\hat{A} \in L(\mathcal{H})$ is an orthogonal projection representing the property a in the Hilbertian quantum logic $L(\mathcal{H})$. Of course numerical values of all probability measures defined on a logic of properties of a quantum system have to remain the same since it makes no difference whether the properties of a system are represented by closed subspaces of a Hilbert space, orthogonal projections onto these subspaces, fuzzy subsets of the unit sphere in a Hilbert space or suitably defined fuzzy subsets of a phase space Γ.

Therefore, the phase space representation of quantum systems outlined above should be free from such counterintuitive ingredients as the negative probabilities which have plagued phase space representations of quantum mechanics from the very birth of this idea. It is our view that the necessity of working with σ-orthomodular posets of fuzzy subsets instead of Boolean σ-algebras of crisp subsets of a phase space is not too high price to be paid for this.

References

1. Ballentine, L. E. "Probability theory in quantum mechanics", *American Journal of Physics*, **54** (1986) 883–889.
2. Santos, E. "The Bell inequalities as tests of classical logics", *Physics letters A*, **115** (1986) 363–365.
3. Pulmannová, S. and V. Majernik, "Bell inequalities on quantum logics", *Journal of Mathematical Physics*, **33** (1992) 2173–2178.
4. Beltrametti, E. G. and M. J. Mczyński, "On the characterization of probabilities: A generalization of Bell's inequalities", *Journal of Mathematical Physics*, **34** (1993) 4919–4929.
5. Beltrametti, E. G. and M. J. Mczyński, "On some probabilistic inequalities related to the Bell inequality", *Reports on Mathematical Physics*, **33** (1993) 123–129.
6. Fine, T. L. *Theories of Probability* (Academic Press, New York, 1973).
7. Zadeh, L. A. "Probability measures on fuzzy events", *Journal of Mathematical Analysis and Applications*, **23** (1968) 421–427.
8. Klement, E. P., R. Lowen, and W. Schwychla, "Fuzzy probability measures", *Fuzzy Sets and Systems*, **5** (1981) 21–30.
9. Piasecki, K. "Probability of fuzzy events as denumerable additivity measure", *Fuzzy Sets and Systems*, **17** (1985) 271–284.
10. Mesiar, R. "Fuzzy sets and probability theory", *Tatra Mountains Mathematical Publications*, **1** (1992) 105–123.
11. Mesiar, R. and M. Navara, "T_s-tribes and T_s-measures", *Journal of Mathematical Analysis and Applications*, **201** (1996) 91–102.
12. Gonseth, F. *Les entretiens de Zürich sur les fondements et la méthode des sciences mathematiques 6–9 décembre 1938* (Zürich, 1941).

Chapter 9
The Many-Valued Interpretation
of Quantum Mechanics

In this chapter we shall present an outline of a proposal expressed in a subtitle of this book: the many-valued interpretation of quantum mechanics, according to the scheme adopted in Chap. 2. However, before we do this, let us consider the following situation.

I go to sleep tonight and I consider to what extent I possess a property $W_7 =$ "being awaken on the next day before 7 a.m.". The degree to which I possess this property at present depends both on my present state $s = \{$my tiredness, the ammount of wine drunk tonight, kind of food eaten for supper, etc.$\}$, and also on the future "experimental arrangement" $e = \{$traffic noise, air temperature, barking of dogs, etc.$\}$. It is obvious that the degree to which I possess the property W_7, or equivalently, the present truth value of the statement "W_7" depends on both s and e.

When we adopt such point of view, it is natural to accept that quantum objects which possess, in the MV sense, all their properties, reveal, depending on experimental arrangements, either wave-like or particle-like or both [1] properties.

In general, numbers from the unit interval traditionally interpreted as probabilities that suitable experiments will reveal some properties of quantum objects, are according to the propounded interpretation reinterpreted as MV truth values or "fuzzy" (i.e., different from 0 or 1) degrees of possessment of these properties. This refers also to superpositions of states. If a state of a quantum object is $|\psi\rangle = \Sigma c_i |\psi_i\rangle$, then the numbers $|c_i|^2$ are traditionally interpreted as probabilities that the object will be found in one of the states $|\psi_i\rangle$ when a suitable experiment is done. According to the propounded interpretation these numbers represent degrees to which the object, which is in the state $|\psi\rangle$, is at the same time in the respective states $|\psi_i\rangle$. Such an interpretation explains, for example, the result of an experiment by Robert et al. [2] (see also [3]), in which an atom in a state that was a superposition of two space-time well separated locations, emitted light exactly as if it was in these two locations simultaneously.

© The Author(s) 2015
J. Pykacz, *Quantum Physics, Fuzzy Sets and Logic*,
SpringerBriefs in Physics, DOI 10.1007/978-3-319-19384-7_9

Main ideas:

- All the mathematical formulation of QM is left intact.
- Quantum mechanics is fundamentally about results of *future* observations or results of *future* experiments.
- Statements about future non-certain events in natural way belong to the domain of many valued logic.
- Quantum objects possess all their properties even before they are measured, however in the MV sense, i.e., to the extent $p \in [0, 1]$, previously interpreted as a probability that suitable measurement would reveal that a property have been possessed.

Virtues:

- Indeterminism.
- No problems with wave-particle duality.
- Disappearance of various "paradoxes" yielded by assumed prior-to-measurement existence (in the sense of 2-valued logic) of properties of quantum objects.
- Full compatibility with the "orthodox quantum logic", i.e., the mathematical structure that is characteristic to a family of closed subspaces of a Hilbert space used in the mathematical description of a quantum system.
- Clarification of the meaning of conjunctions and disjunctions of statements about quantum objects.
- Opening the possibility of constructing quantum probability calculus in a full analogy to the classical Kolmogorovian probability calculus.
- Opening the possibility of constructing (fuzzy) phase-space representation of quantum mechanics.

Drawbacks:

Before we state two obvious drawbacks of the proposed interpretation we would like to draw attention of a reader to the fact that this interpretation of QM is still "in statu nascendi". Therefore, we do hope that these drawbacks may disappear in the future.

- Inability to explain the apparent existence of non-local correlations between properties of spatially separated objects.[1]
- Inability to solve the "objectification problem", i.e., a problem how "potential" properties become "actual" in the course of a measurement.[2]

---

[1] This is not a problem to a vast number of scholars that are comfortable with the apparent "non-locality of QM". Our "guts feeling" is that no influence, carrying information or not, should propagate faster than light. This issue is, however, not addressed at the present stage of development of the proposed MVI of QM.

[2] This problem is solved if MVI is applied to any of "Objective Collapse Theories", but of course not when it is applied to the "Orthodox QM".

References

1. Mittelstaedt, P., A. Prieur, and R. Schieder, "Unsharp particle-wave duality in a photon split-beam experiment", *Foundations of Physics*, **17** (1987) 891–903.
2. Robert, J. et al. "Atomic quantum phase studies with a longitudinal Stern-Gerlach interferometer". *Journal de Physique II*, **2** (1992) 601–614.
3. Czachor, M. and L. You, "Spatially sequential turn-on of spontaneous emission from an atomic wave packet", *International Journal of Theoretical Physics*, **38** (1999) 277–288.

Index

© The Author(s) 2015
J. Pykacz, *Quantum Physics, Fuzzy Sets and Logic*,
SpringerBriefs in Physics, DOI 10.1007/978-3-319-19384-7

Printed in the United States
By Bookmasters